Machine Learning-Based Fault Diagnosis for Industrial Engineering Systems

Advances in Intelligent Decision-Making, Systems Engineering, and Project Management

This new book series will report the latest research and developments in the field of information technology, engineering and manufacturing, construction, consulting, healthcare, military applications, production, networks, traffic management, crisis response, human interfaces, and other related and applied fields. It will cover all project types, such as organizational development, strategy, product development, engineer-to-order manufacturing, infrastructure and systems delivery, and industries and industry sectors where projects take place, such as professional services, and the public sector including international development and cooperation, etc. This new series will publish research on all fields of information technology, engineering, and manufacturing including the growth and testing of new computational methods, the management and analysis of different types of data, and the implementation of novel engineering applications in all areas of information technology and engineering. It will also publish on inventive treatment methodologies, diagnosis tools and techniques, and the best practices for managers, practitioners, and consultants in a wide range of organizations and fields including police, defense, procurement, communications, transport, management, electrical, electronic, aerospace, requirements.

Blockchain Technology for Data Privacy Management
Edited by Sudhir Kumar Sharma, Bharat Bhushan, Aditya Khamparia, Parma Nand Astya, and Narayan C. Debnath

Smart Sensor Networks Using AI for Industry 4.0
Applications and New Opportunities
Edited by Soumya Ranjan Nayak, Biswa Mohan Sahoo, Muthukumaran Malarvel, and Jibitesh Mishra

Hybrid Intelligence for Smart Grid Systems
Edited by Seelam VSV Prabhu Deva Kumar, Shyam Akashe, Hee-Je Kim, and Chinmay Chakrabarty

Machine Learning-Based Fault Diagnosis for Industrial Engineering Systems
Rui Yang and Maiying Zhong

For more information about this series, please visit: https://www.routledge.com/Advances-in-Intelligent-Decision-Making-Systems-Engineering-and-Project-Management/book-series/CRCAIDMSEPM

Machine Learning-Based Fault Diagnosis for Industrial Engineering Systems

Rui Yang
Maiying Zhong

CRC Press
Taylor & Francis Group
Boca Raton London New York

CRC Press is an imprint of the
Taylor & Francis Group, an **informa** business

First edition published 2022
by CRC Press
6000 Broken Sound Parkway NW, Suite 300, Boca Raton, FL 33487-2742

and by CRC Press
4 Park Square, Milton Park, Abingdon, Oxon, OX14 4RN

CRC Press is an imprint of Taylor & Francis Group, LLC

© 2022 Rui Yang, Maiying Zhong

Reasonable efforts have been made to publish reliable data and information, but the author and publisher cannot assume responsibility for the validity of all materials or the consequences of their use. The authors and publishers have attempted to trace the copyright holders of all material reproduced in this publication and apologize to copyright holders if permission to publish in this form has not been obtained. If any copyright material has not been acknowledged please write and let us know so we may rectify in any future reprint.

Except as permitted under U.S. Copyright Law, no part of this book may be reprinted, reproduced, transmitted, or utilized in any form by any electronic, mechanical, or other means, now known or hereafter invented, including photocopying, microfilming, and recording, or in any information storage or retrieval system, without written permission from the publishers.

For permission to photocopy or use material electronically from this work, access www.copyright.com or contact the Copyright Clearance Center, Inc. (CCC), 222 Rosewood Drive, Danvers, MA 01923, 978-750-8400. For works that are not available on CCC please contact mpkbookspermissions@tandf.co.uk

Trademark notice: Product or corporate names may be trademarks or registered trademarks and are used only for identification and explanation without intent to infringe.

Library of Congress Cataloging-in-Publication Data
Names: Yang, Rui (Professor of computer engineering), author. | Zhong,
   Maiying, author.
Title: Machine learning-based fault diagnosis for industrial engineering
   systems / Rui Yang, Maiying Zhong.
Description: First edition. | Boca Raton : CRC Press, 2022. | Series:
   Advances in intelligent decision-making | Includes bibliographical
   references and index.
Identifiers: LCCN 2021060649 (print) | LCCN 2021060650 (ebook) | ISBN
   9781032147253 (hardback) | ISBN 9781032147260 (paperback) | ISBN
   9781003240754 (ebook)
Subjects: LCSH: Fault location (Engineering)--Automation. | Automatic test
   equipment. | Machinery--Testing. | Industrial equipment--Maintenance and
   repair. | Machine learning.
Classification: LCC TA169.6 .Y365 2022 (print) | LCC TA169.6 (ebook) |
   DDC 620/.0044--dc23/eng/20220325
LC record available at https://lccn.loc.gov/2021060649
LC ebook record available at https://lccn.loc.gov/2021060650

ISBN: 978-1-032-14725-3 (hbk)
ISBN: 978-1-032-14726-0 (pbk)
ISBN: 978-1-003-24075-4 (ebk)

DOI: 10.1201/9781003240754

Typeset in Times
by SPi Technologies India Pvt Ltd (Straive)

Contents

Preface ix
Authors xi

Chapter 1 Background and Related Methods 1

    1.1   Background...1
    1.2   Related Methods 2
            1.2.1   Back Propagation Neural Network.............................2
            1.2.2   Convolutional Neural Network 3
            1.2.3   Recurrent Neural Network 4
            1.2.4   Generative Adversarial Networks...............................5
            1.2.5   Bagging Algorithm...5
            1.2.6   Classification and Regression Tree............................6
            1.2.7   Random Forest 6
            1.2.8   Density-Based Spatial Clustering of Applications with Noise 7
            1.2.9   Safe-Level Synthetic Minority Over-Sampling Technique 8
    Bibliography...8

Chapter 2 Fault Diagnosis Method Based on Recurrent Convolutional Neural Network 11

    2.1   Introduction 11
    2.2   Model Establishment and Theoretical Derivation 11
            2.2.1   One-Dimensional Convolutional Neural Network 12
            2.2.2   Convolutional Recurrent Neural Network Model 13
            2.2.3   Dropout in Neural Network Model 18
    2.3   Diagnostic Flow of the Proposed Method 19
    2.4   Experimental Research Based on The Proposed Method........20
            2.4.1   Experiment Platform 20
            2.4.2   Experimental Setup 21
            2.4.3   Summary of Experimental Results............................22
    Bibliography...23

Chapter 3 Fault Diagnosis of Rotating Machinery Gear Based on Random Forest Algorithm 27

    3.1   Introduction 27
    3.2   Fault Diagnosis of Rotating Machinery Gear Based on Random Forest Algorithm 28

		3.3	Experimental Verification 30

3.3 Experimental Verification 30
    3.3.1 Experiment Platform 31
    3.3.2 Experimental Results..33
    3.3.3 Comparison Study 35
Bibliography..37

Chapter 4   Bearing Fault Diagnosis under Different Working Conditions Based on Generative Adversarial Networks 39

4.1 Introduction 39
4.2 Model Establishment and Theoretical Derivation 40
    4.2.1 Wasserstein Generative Adversarial Network 40
    4.2.2 Maximum Mean Discrepancy 41
    4.2.3 Establishment of Fault Diagnosis Model 42
    4.2.4 Fault Diagnosis Procedures of the Proposed Method 43
4.3 Experimental Results...44
Bibliography..45

Chapter 5   Rotating Machinery Gearbox Fault Diagnosis Based on One-Dimensional Convolutional Neural Network and Random Forest.....47

5.1 Introduction 47
5.2 Model Establishment and Theoretical Derivation 48
    5.2.1 One-Dimensional Convolutional Neural Network 49
    5.2.2 Random Forest Algorithm 50
    5.2.3 The Proposed Fault Diagnosis Model 51
    5.2.4 Error Back Propagation of the Proposed Model 51
    5.2.5 Weights Optimization Using Adaptive Moments........54
5.3 Experimental Results...54
    5.3.1 Experimental Platform 54
    5.3.2 Experimental Setup 55
    5.3.3 Analysis of Experimental Results 55
Bibliography..57

Chapter 6   Fault Diagnosis for Rotating Machinery Gearbox Based on Improved Random Forest Algorithm..59

6.1 Introduction 59
6.2 Improved Random Forest Algorithm..60
    6.2.1 Semi-Supervised Learning 60
    6.2.2 Improved Random Forest Classification Algorithm 62
6.3 Experimental Verification 63
Bibliography..64

Chapter 7		mbalanced Data Fault Diagnosis Based on Hybrid Feature Dimensionality Reduction and Varied Density-Based Safe-Level Synthetic Minority Oversampling Technique 67	
	7.1	Introduction 67	
	7.2	Design of Hybrid Feature Dimensionality Reduction Algorithm 68	
		7.2.1	Sensitive Feature Selection.. 69
		7.2.2	Dimension Reduction of Features 70
	7.3	Design of Varied Density-Based Safe-Level Synthetic Minority Oversampling Technique... 71	
	7.4	Experiment and Results.. 72	
		7.4.1	Data Classification Method 72
		7.4.2	Experiment Platform 74
		7.4.3	Feature Extraction 74
		7.4.4	Data Acquisition 75
		7.4.5	Results Analysis 76
	Bibliography... 77		

Index... 79

Preface

Industrial mechanical equipment is an indispensable and essential part of industrial production. Intelligent equipment is increasingly becoming the direction of industrial system development due to the development of intelligent technologies. With the improved requirement of the safety and reliability of industrial equipment, the research on equipment fault diagnosis is becoming critical. Rotating machinery, as a typical mechanical equipment that relies on rotation to complete specific functions, plays an essential role in industrialization. Various types of rotating machinery such as water turbines, steam turbines, gas turbines, wind turbines, and aero engines are widely used in industrial fields such as electricity, petrochemical, and aerospace. Because rotating machinery is usually working in a harsh environment with long working hours, the condition of the rotating machinery must be monitored to detect the problems in time to avoid potential accidents. Therefore, the research on various fault diagnosis problems in rotating machinery has become increasingly important. Although the research on fault diagnosis of rotating machinery has achieved rich results, many problems still need to be solved, such as the problem in automatic knowledge acquisition, the difficulty in mathematical modeling, and the insufficient multi-fault diagnosis.

In recent years, mechanical equipment has developed in complex and intelligent directions. The research on fault diagnosis that realizes intelligent diagnosis with multiple faults and multiple working conditions has attracted more and more attention. As the promising development directions of machine learning, neural network and deep learning play important roles in automatic reasoning and cognitive modeling. Considering the structure of rotating machinery and the characteristics of neural networks, the application of machine learning in fault diagnosis of rotating machinery can fully use the advantages of machine learning in data processing and category identification with a high chance of finding a breakthrough in mechanical fault diagnosis. In summary, rotating machinery is widely used in industry and has a prominent role in economic development. The machine learning methods are suitable for handling complex mechanical fault diagnosis problems due to the advantages in intelligent data analysis. Therefore, the machine learning-based fault diagnosis of industrial equipment in this book is a promising research direction with high practical value. The existing fault diagnosis abilities of existing methods are enhanced by improving and combining conventional machine learning methods. This book can provide technical support for scientific researchers and postgraduate students in related fields.

There are seven chapters in this book. Chapter 1 introduces the background of this book and provides the general concepts of the related machine learning methods. Chapters 2 and 3 focus on fault diagnosis using single machine learning methods. Chapter 4 investigates the fault diagnosis problem under multiple working conditions. Chapters 5 and 6 focus on fault diagnosis using hybrid and improved machine learning methods. Chapter 7 investigates the fault diagnosis problem under imbalanced data condition.

The research works of this book were partially supported by National Natural Science Foundation of China, Research Fund for the Taishan Scholar Project of Shandong Province of China, Jiangsu Provincial Qinglan Project, and Research Development Fund of Xi'an Jiaotong-Liverpool University.

Authors

Rui Yang received the B.Eng. degree in Computer Engineering and the Ph.D. degree in Electrical and Computer Engineering from National University of Singapore in 2008 and 2013 respectively. He is currently an assistant professor in the School of Advanced Technology, Xi'an Jiaotong-Liverpool University, Suzhou, China, and an honorary lecturer in the Department of Computer Science, University of Liverpool, Liverpool, United Kingdom. His research interests include machine learning-based data analysis and applications.

Maiying Zhong received her Ph.D. degree in control theory and control engineering from the Northeastern University, China, in 1999. From 2000 to 2001, she was a visiting scholar at the University of Applied Sciences Lausitz, Germany. From 2002 to July 2008, she was a professor at the School of Control Science and Engineering at Shandong University. From 2006 to 2007, she was a postdoctoral research fellow with the Central Queensland University, Australia. From 2009 to 2016, she was a professor at the School of Instrument Science and Opto-Electronics Engineering, Beihang University. In March 2016, she joined Shandong University of Science and Technology, China, where she is currently a professor with the College of Electrical Engineering and Automation. Her research interests are model-based fault diagnosis, fault tolerant systems, and their applications.

1 Background and Related Methods

1.1 BACKGROUND

For the machinery in modern industry, the complex structure, large-scale design, and intelligent functions are the new development directions. Due to the influence of many factors during operation such as heavy load and high temperature, the core components of mechanical equipment such as gears and bearings may experience certain degradation, fault, or even completely failure. The failure of mechanical components can cause a series of problems, ranging from property damage to human severe injury. Therefore, it is an indispensable part of modern industry to apply fault diagnosis technology to perform real-time monitoring during the operation of equipment, detect the fault as early as possible, and provide a reliable basis for equipment maintenance.

Although the fault diagnosis of industrial equipment has a long history and has accumulated a lot of experience, mechanical equipment has been developing fast in highly complex and intelligent directions in recent years. Therefore, intelligent fault diagnosis research for multiple faults and multiple working conditions has attracted the attention of researchers. Modern industrial equipment is large and complex with many monitoring points and sensors, which bring a large amount of data and information for analysis and recognition. The big data of mechanical equipment brings new challenges to fault diagnosis, which can be summarized as the following two points:

(1) The manual detection of fault is very difficult to meet the requirement with a large amount of data, which requires automatic and intelligent fault diagnosis algorithm.
(2) The data types are diversified, and each sample may be obtained from different machines at different positions under different working conditions, which increase the difficulty of feature mining and fault diagnosis.

Because of the above two challenges, using massive data for feature mining to achieve efficient and accurate fault diagnosis is a complex problem. In recent years, the development of artificial intelligence technology has been relatively rapid. As important disciplines and promising development directions of artificial intelligence, neural networks, and deep learning theories have developed fast and show the importance in directions such as automatic reasoning, cognitive modeling, and intelligent manufacturing. At the same time, the intelligent fault diagnosis method of mechanical equipment is also thriving, which combines fault diagnosis theory and machine learning methods. The intelligent fault diagnosis methods perform feature mining on a large number of signals extracted by various sensors to obtain feature information reflecting the fault, establish the diagnosis model through the analysis of various feature information, and conduct accurate and reliable fault diagnosis in mechanical equipment.

In summary, the combination of machine learning and fault diagnosis can make full use of the massive amounts of data to diagnose faults on industrial machinery without much human involvement. Therefore, the machine learning-based fault diagnosis of mechanical equipment has essential research significance and broad research prospects.

1.2 RELATED METHODS

1.2.1 BACK PROPAGATION NEURAL NETWORK

Back propagation neural network is a network formed by interconnecting multiple layers of neurons. It has a very powerful nonlinear mapping capability and is the basis for many complex network structures. The loss function and the gradient descent method are the core parts of the back propagation neural network. The value of the neuron node in each layer can be calculated as follows:

$$\begin{cases} y^{(l)}(j) = \sum_{i=1}^{I} w_{ji}^{(l)} x^{(l-1)}(i) \\ x^{(l)}(j) = f\left(y^{(l)}(j)\right) \end{cases} \tag{1.1}$$

where $x^{(l)}(j)$ represents the activation value of the jth neuron in the lth layer, $w^{(l)}_{ji}$ represents the weight between the ith neuron in the l-1th layer, and the jth neuron in the lth layer and f refers to the activation function. Different activation functions can be designed for different problems in practical applications according to the requirements.

Usually, the loss function of the back propagation neural network is represented by the sum of variances between the actual value and the expected value, and the computation is as follows:

$$\text{loss} = \frac{1}{2} \sum_{n=1}^{N} \left(o(n) - d(n)\right)^2 \tag{1.2}$$

where o is the actual output value and d is the expected value.

Then the partial derivative of the loss function to the weights of the last layer and the neurons in the previous layer can be computed as:

$$\begin{cases} \dfrac{\partial \text{loss}}{\partial w_n^{(L)}} = -\left(o(n) - d(n)\right) f'\left(y^{(L)}(n)\right) x^{(L-1)} \\ \dfrac{\partial \text{loss}}{\partial x^{(L-1)}(m)} = -\sum_{n=1}^{N} \left(o(n) - d(n)\right) f'\left(y^{(L)}(n)\right) w_{nm}^{(L)} \end{cases} \tag{1.3}$$

where $w_n^{(L)}$ represents the weight vector of the neuron of the previous layer to the nth neuron of the Lth layer, and f represents the differential form of the activation function. By analogy, the weight gradient and neuron gradient of the entire network can be obtained.

1.2.2 CONVOLUTIONAL NEURAL NETWORK

Convolutional neural network is a multi-level network composed of multiple convolutional layers, and the idea comes from the visual cortex classification principle of the biological nervous system. The basic structure of convolutional neural network includes four essential layers: convolutional layer, pooling layer, activation layer, and fully connected layer. As a classic model in deep learning theory, the convolutional neural network adopts the method of sparse local connection and weight sharing, which avoids the problem of intensive weight calculations in ordinary multi-layer networks due to full connections. Convolutional neural networks have been applied to machine vision, speech recognition, fault diagnosis, and other fields in recent years and have achieved great success in these fields.

The convolutional layer is the core of the convolutional neural network. It is characterized by the ability to share weights among neurons and the primary purpose is to extract data features. It is also a vital part of the convolutional neural network that is different from the ordinary neural network. The convolutional layer performs convolution operations on the local area of the input signal (or the feature vector provided by the previous layer) through a specific size of convolution kernel and obtains the data features. For the sparse local connection, each neuron in each layer only perceives part of the input data. For the weight sharing, the neurons use the same convolution kernel to compute the outputs. As the most essential feature of the convolutional layer, weight sharing can reduce network parameters to reduce the phenomenon of overfitting.

To increase the representation ability of the network, it is necessary to perform activation operation on these features and map the features extracted from convolution operation to a linearly separable feature space. The activation functions commonly used in neural networks include the Sigmoid, Tanh (hyperbolic tangent), and ReLU (rectified linear unit).

In the convolutional neural network, the pooling operation is a down-sampling process, and the primary purpose is to reduce the amount of calculation by reducing the size of the feature map, which is usually used after the activation layer. The pooling process is similar to a sliding window through which a specific area of the input data is pooled. The most commonly used pooling techniques are maximum pooling and average pooling. The maximum pooling is to output the maximum value in each area, and the average pooling is to output the average value in each area.

In the entire convolutional neural network, the convolutional layer, activation layer, and pooling layer are used as feature extraction modules. Those features extracted by the feature extraction module are usually local, and the role of the fully connected layer is to integrate these local features for classification and is generally the last layer of the neural network.

In summary, the convolutional neural networks have the advantages of sharing convolution kernels with a local perceptive field and have strong data analysis capability for high-dimensional features. Training an excellent convolutional neural network model usually requires efforts in parameter tuning and large number of samples.

1.2.3 Recurrent Neural Network

An important idea in deep learning theory is parameter sharing. In addition to the convolution kernel sharing used in convolutional neural networks, the layer weight sharing in recurrent neural networks is also a well-applied method in deep learning. The recurrent neural network model was first proposed in 1986. Compared with the multi-layer fully connected network, the recurrent neural network has simple recurrent units and shared weights and can be extended to sequential analysis with different data lengths. A simple RNN network is shown in Figure 1.1, which shows three basic cyclic units.

From Figure 1.1, the weight matrix V, W, and U form a network in a fully connected form, and the output o of each unit can be selected as the eigenvector of the network according to the actual situation. Assuming that the numbers of neurons in the three layers of x, h, and o are J, K, and I respectively, the computation process of the network is:

$$\begin{cases} h_i^{(n)} = \tanh\left(\sum_{j=1}^{J} w_{ij} x_j^{(n)} + \sum_{k=1}^{K} u_{ik} h_k^{(n-1)}\right) \\ o_l^{(n)} = \tanh\left(\sum_{i=1}^{I} v_{li} h_i^{(n)}\right) \end{cases} \quad (1.4)$$

where $h_i^{(n)}$ represents the activation value of the i-th neuron in the hidden layer of the nth unit, and w_{ij} and u_{li} respectively represent the weight of the neuron at the corresponding position.

The recurrent neural network can perform efficient recurrent learning with nonlinear sequential data. With weight sharing, the recurrent neural network solves the

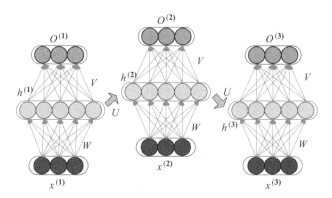

FIGURE 1.1 Illustration of recurrent neural network structure.

massive parameter training defect of multi-layer network. Overall, recurrent neural networks have shown exemplary performance in speech recognition, language modeling, machine translation and other fields. At the same time, they are also widely used in time series forecasting or computer vision combined with other types of networks.

1.2.4 GENERATIVE ADVERSARIAL NETWORKS

The generative adversarial networks are based on machine learning and game theory, and the main idea is to recognize the generated data from the original data until the generated data is recognized as real data. The generative adversarial network is a generative model based on adversarial ideas, consisting of two parts: generator (G) and discriminator (D). Generative adversarial network completes the task through the generator to generate data that can be identified as "true" by the discriminator through multiple iterations of training. First, the generator generates data through random noise, and then inputs these data into the discriminator. Then the discriminator uses the real input data as the basis for identification of the input generated data. If the identification is "true", it indicates that the generated data has been considering similar to the real data. If the identification is "false", the discriminator will feed back the result to the generator and instruct the generator to generate data again. The idea of generative and adversary has been applied to the field of transfer learning in recent years to reduce the difference between the feature distributions of two domains.

The training mechanism for generative adversarial networks is as follows:

(1) At the initial stage, the discriminator cannot distinguish between real data and generated data.
(2) The discriminator is optimized so that the discriminator has specific discriminating ability, under the condition that the generator remains unchanged.
(3) The generator is also optimized according to the gradient information from the discriminator to generate data that the discriminator cannot distinguish, to reduce the distribution difference of the generated data and the real data.
(4) After multiple updates and optimizations, the distributions of the generated and real data are consistent, and the discriminator cannot discriminate between the generated and real data.

1.2.5 BAGGING ALGORITHM

Bagging algorithm is a random data selection method proposed by Leo Breiman in 1996. This algorithm uses the bootstrap re-sampling method to randomly sample with replacement for k times from the original data set with a capacity of n, to form k bootstrap sub-sample sets. The sample size drawn each time is equal to the sample size of the original data set, and k sub-sample sets are used to train k classifiers. It has been proved that this algorithm can improve the generalization ability of each classifier.

After k times random sampling, the probability of data in the original data set which is not selected is $\left(1-\frac{1}{k}\right)^k$. The selected data are used as the training data set, and the unselected data are called out-of-bag data and are used as the test data set. The out-of-bag error can be used to estimate the classification accuracy of other classification algorithms.

1.2.6 Classification and Regression Tree

The classification and regression tree algorithm is a two-classification algorithm. In the tree growth process, each non-leaf node uses specific rules to select the optimal split feature to divide the sample set into two subsets until all the features are used. The optimal split feature and the optimal split value are selected according to the principle of the smallest Gini index. Assuming that the current feature contains C categories, the Gini value can be computed as below:

$$Gini(t_i) = \sum_{j=1}^{C}\sum_{j'\neq j} p_j p_{j'} = 1 - \sum_{j=1}^{C} p_j^2 \tag{1.5}$$

where p_j is the probability of the occurrence of the j-th category. It can be seen that the smaller the Gini value is, the higher the purity of the category is. Therefore, the feature with the smallest Gini value traversing all the features of the tree should be used as the split feature of the current node.

Assuming that the optimal split feature is t_i, the optimal split value a can be selected as below:

$$\min_{a} Gini(t_i, a) = \frac{N_1}{N} Gini(t_{i_1}) + \frac{N_2}{N} Gini(t_{i_2}) \tag{1.6}$$

where t_{i_1} and t_{i_2} are the two sub-sample sets of the optimal split feature obtained by enumeration, N_1 and N_2 are the number of samples in the two sub-sample sets, and N is the number of samples with the optimal split feature t_i.

The growth of the decision tree is completed when all m features are used to split nodes. The classification result of the tree is obtained by taking the majority outcome after synthesizing the prediction results of each leaf node.

1.2.7 Random Forest

Random forest is an algorithm based on the feature set. As the quality of features restricts the classification accuracy of the algorithm to a certain extent, the selection of features is of great significance in random forests. Assuming that the random forest is composed of k classifiers $h_1(T), h_2(T), \cdots, h_k(T)$, it can be expressed as:

$$H = \{h_1(T), h_2(T), \cdots, h_k(T)\} \tag{1.7}$$

Background and Related Methods

where T is the input feature set and $h_\theta(T)$, $\theta = 1, 2, \cdots, k$ is the base classifier. Each classifier is a decision tree generated by the classification and regression tree algorithm. The classification results of k decision trees are voted and the majority result is the classification result of the random forest.

The implementation process of the random forest algorithm includes the following parts:

(1) Bootstrap re-sampling
    Select from the original data set randomly with replacement k times using the bootstrap re-sampling method (where k is the number of decision trees in the random forest) to form k bootstrap sub-sample sets, and each sub-sample set contains n samples (n is the number of samples in the original data set).
(2) Growth of classification and regression tree
    According to the Bagging algorithm and the random subspace method, m features are randomly selected as a feature subset. The feature subset is used as the division attribute in the growth process of the classification and regression tree according to the principle of the smallest Gini value. The selection of random subspace reduces the correlation of each decision tree in the random forest. Therefore, the prediction accuracy of the entire forest can be improved by combining the classification results of each tree.
(3) Construction of random forest
    According to the method introduced in step (2) above, use k bootstrap sub-sample sets to grow k decision trees to form a random forest.
(4) Classification results of random forest
    Synthesize the classification results of k decision trees and use the voted result as the final classification result of the random forest.

1.2.8 Density-Based Spatial Clustering of Applications with Noise

Density-based spatial clustering of applications with noise is a density-based clustering algorithm, which implements clustering of data through parameters *Eps* and *MinPts*, where *Eps* represents the neighborhood distance threshold of a specific sample, and *MinPts* represents the minimum number of samples required to form a cluster. This clustering algorithm can form arbitrary-shaped clusters and filter out noise data.

The appropriate value of the parameter *MinPts* is usually selected according to the principle of $(M + 1) \leq MinPts \leq 2M$, where M represents the feature dimension of the sample. The value of *Eps* is generally obtained through the sorted k-distance curve and picking the value corresponding to the change point in the k-distance curve.

Assuming a data set $D = (x_1, x_2, \ldots, x_N)$, the Euclidean distance $dist(x_i, x_j)$ between any sample x_i in the data set D and its k-th nearest neighbor x_j can be calculated. This distance $dist(x_i, x_j)$ is called k-distance. Sort the k-distance value from small to large and put the sorted k-distance into the *kdist* set, then the curve drawn according to the distance value in *kdist* is called the sorted k-distance curve.

The density-based spatial clustering of applications with noise algorithm realizes the clustering of the data set D by accessing each sample x and $N_{Eps}(x)$. In this algorithm, if x is the core point, a new cluster should be created and all samples in the cluster $N_{Eps}(x)$ into the newly created cluster. Then, visit all the samples in the new cluster, find out all the core points, and add all the directly density-reachable samples from the core point to the cluster until no new samples can be found. Then the clustering process of one cluster is completed. The algorithm will revisit the data that has never been visited before, and if the conditions of the above process are met, a new cluster will be created again.

1.2.9 Safe-Level Synthetic Minority Over-Sampling Technique

Safe-level synthetic minority over-sampling technique is proposed to solve the problems of boundary overlap and noise in imbalanced data classification. The algorithm defines a parameter sl (safety level) for each minority sample as the number of any minority sample in its K-nearest neighbor range. Another parameter, sl_ratio, is defined as the ratio of the safety level of the minority sample and its selected neighbor samples of the same class. A new sample is synthesized by calculating the sl of each minority sample and one of its K nearest neighbors and by using sl_ratio as the weight of the linear interpolation between them. The following equation shows the linear interpolation method for synthesizing new minority samples.

$$s' = s + w * (q - s) \tag{1.8}$$

where s' represents a synthesized sample, s represents a minority sample, q represents a minority sample of its neighbor s, and w represents a linear interpolation weight determined by sl_ratio.

In the safe-level synthetic minority over-sampling algorithm, the value range of w is determined by the value of sl_ratio, which makes the synthetic data tend to be generated near samples with higher safety levels. This design can ensure the safety and effectiveness of the synthetic data and overcome the problems caused by conventional synthetic minority over-sampling algorithm. According to the value of sl_ratio, the ways of data synthesis are divided into the following five situations:

1) If $sl\_ratio = \infty$ and $sl_s = 0$, no new samples are generated between s and q;
2) If $sl\_ratio = \infty$ and $sl_s \neq 0$, then q is noise data, and new data is generated by copying s;
3) If $sl\_ratio = 1$, the synthetic data is generated in the range of [0, 1];
4) If $sl\_ratio > 1$, the synthetic data is generated in the range of [0, $1/sl\_ratio$];
5) If $sl\_ratio < 1$, the synthetic data is generated in the range of [$1-sl\_ratio$, 1].

BIBLIOGRAPHY

Arjovsky, M., Chintala, S. & Bottou, L. Wasserstein generative adversarial networks. *International Conference on Machine Learning*, 2017. 214–223.

Bi, F.-M., Wang, W.-K. & Chen, L. 2012. DBSCAN: Density-based spatial clustering of applications with nose. *Journal of NanJing University (Natural Sciences)*, 48, 491–498.

Breiman, L. 1996. Bagging predictors. *Machine Learning*, 24, 123–140.

Breiman, L. 2001. Random forests. *Machine Learning*, 45, 5–32.

Bunkhumpornpat, C., Sinapiromsaran, K. & Lursinsap, C. Safe-level-smote: Safe-level-synthetic minority over-sampling technique for handling the class imbalanced problem. *Pacific-Asia Conference on Knowledge Discovery and Data Mining*, 2009. Springer, 475–482.

Chandel, A. K. & Patel, R. K. 2013. Bearing fault classification based on wavelet transform and artificial neural network. *IETE Journal of Research*, 59, 219–225.

Cutler, A., Cutler, D. R. & Stevens, J. R. 2012. Random forests. *Ensemble Machine Learning*. Springer.

Da Silva, W., Habermann, M., Shiguemori, E. H., Do Livramento Andrade, L. & De Castro, R. M. Multispectral image classification using multilayer perceptron and principal components analysis. *2013 BRICS Congress on Computational Intelligence and 11th Brazilian Congress on Computational Intelligence*, 2013. IEEE, 557–562.

Ester, M., Kriegel, H.-P., Sander, J. & Xu, X. A density-based algorithm for discovering clusters in large spatial databases with noise. *Second International Conference on Knowledge Discovery and Data Mining*, 1996. 226–231.

Esteva, A., Kuprel, B., Novoa, R. A., Ko, J., Swetter, S. M., Blau, H. M. & Thrun, S. 2017. Dermatologist-level classification of skin cancer with deep neural networks. *Nature*, 542, 115–118.

Goodfellow, I., Pouget-Abadie, J., Mirza, M., Xu, B., Warde-Farley, D., Ozair, S., Courville, A. & Bengio, Y. 2014. Generative adversarial nets. *Advances in Neural Information Processing Systems*, 27, 1–9.

Gulrajani, I., Ahmed, F., Arjovsky, M., Dumoulin, V. & Courville, A. 2017. Improved training of Wasserstein GANs. *arXiv preprint arXiv:1704.00028*.

He, K., Zhang, X., Ren, S. & Sun, J. Deep residual learning for image recognition. *Proceedings of the IEEE Conference on Computer Vision and Pattern Recognition*, 2016. 770–778.

Hjelm, R. D., Jacob, A. P., Che, T., Trischler, A., Cho, K. & Bengio, Y. 2017. Boundary-seeking generative adversarial networks. *arXiv preprint arXiv:1702.08431*.

Ho, T. K. Random decision forest. *Proceedings of the 3rd International Conference on Document Analysis and Recognition*, 1995. 278282.

Ho, T. K. 1998. The random subspace method for constructing decision forests. *IEEE Transactions on Pattern Analysis and Machine Intelligence*, 20, 832–844.

Krizhevsky, A., Sutskever, I. & Hinton, G. E. 2012. Imagenet classification with deep convolutional neural networks. *Advances in Neural Information Processing Systems*, 25, 1097–1105.

Lecun, Y., Bottou, L., Bengio, Y. & Haffner, P. 1998. Gradient-based learning applied to document recognition. *Proceedings of the IEEE*, 86, 2278–2324.

Masetic, Z. & Subasi, A. 2016. Congestive heart failure detection using random forest classifier. *Computer Methods and Programs in Biomedicine*, 130, 54–64.

Sadeghi, S. & Ramanathan, K. A Hubel Wiesel model of early concept generalization based on local correlation of input features. *The 2011 International Joint Conference on Neural Networks*, 2011. IEEE, 709–716.

Schapire, R. E. 1990. The strength of weak learnability. *Machine Learning*, 5, 197–227.

Silver, D., Huang, A., Maddison, C. J., Guez, A., Sifre, L., Van Den Driessche, G., Schrittwieser, J., Antonoglou, I., Panneershelvam, V. & Lanctot, M. 2016. Mastering the game of Go with deep neural networks and tree search. *Nature*, 529, 484–489.

Strobl, C., Malley, J. & Tutz, G. 2009. An introduction to recursive partitioning: rationale, application, and characteristics of classification and regression trees, bagging, and random forests. *Psychological Methods*, 14, 323–348.

Zeng, W., Xu, C., Zhao, G., Wu, J. & Huang, J. 2018. Estimation of sunflower seed yield using partial least squares regression and artificial neural network models. *Pedosphere*, 28, 764–774.

Zhuang, J.-F., Luo, J., Peng, Y.-Q., Huang, C.-Q. & Wu, C.-Q. 2009. Fault diagnosis method based on modified random forests. *Computer Integrated Manufacturing Systems*, 15, 777–785.

2 Fault Diagnosis Method Based on Recurrent Convolutional Neural Network

2.1 INTRODUCTION

This chapter combines the feature extraction and classification in the conventional fault diagnosis method into a deep learning network, which simplifies the overall operation steps in the fault diagnosis of industrial engineering systems. Because the working environment of rotating machinery is complex and changeable, the types and sizes of faults are not easy to characterize or effectively classify with external noise or disturbances. The method in this chapter uses a combination of convolutional neural networks and recurrent neural networks, which is the basis for obtaining vibration signals of rotating machinery.

In this chapter, the convolutional neural network is used to analyze the internal relationship of the sampled data to receive the main features, and the recurrent neural network is used to summarize and compare the feature extraction results of multiple adjacent periods to obtain a deeper information representation. Thus, a convolutional recurrent neural network fault diagnosis model with deep analysis and internal relationship characterization abilities is proposed. The convolutional recurrent neural network-based bearing fault diagnosis method can avoid cumbersome feature extraction, feature selection, feature fusion, all of which require manual adjustments in fault diagnosis. Compared with conventional deep learning methods, the proposed method has the advantage of high recognition accuracy. The experimental results show that the proposed method can complete the classification task and fault diagnosis target of multiple types of bearing faults under multiple operating conditions.

2.2 MODEL ESTABLISHMENT AND THEORETICAL DERIVATION

The convolutional recurrent neural network model proposed in this chapter is mainly composed of two types of classic deep learning networks: convolutional neural network and recurrent neural network. After combination, a fault diagnosis network system with in-depth analysis capabilities in both vertical and horizontal aspects is formed. It has the following advantages:

1) Using convolutional neural network and recurrent neural network to directly analyze time domain signals and extract depth information, avoiding the complex process of constructing feature vectors using various feature selection methods;
2) The proposed model has the ability to mine internal relationships and conduct in-depth analysis in the time frame. With a relatively simple structure, the proposed model can handle the identification of multiple working conditions and multiple fault categories with low network depth and high diagnostic accuracy.

2.2.1 One Dimensional Convolutional Neural Network

Convolutional neural networks usually have good two dimensional image recognition abilities and can perform in-depth analysis and information mining on the two-dimensional matrix. The multi-condition bearing vibration signals collected are usually stored in the form of vectors, which can be considered as one-dimensional images as a special case. The one-dimensional convolutional neural network model is shown in Figure 2.1.

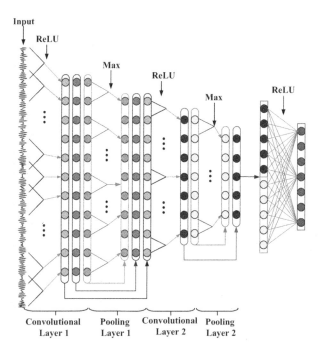

FIGURE 2.1 Structure of one-dimensional CNN.

Convolutional neural network can extract the deep features of the original signal with noise and express the features with a one-dimensional feature vector. The convolution operation method is:

$$y_{\theta z}^{(m)}(r) = \sum_{s=1}^{S} C_s^{(m)}(r) p_{zs}^{(m-1)}(\theta) \tag{2.1}$$

where, $y_{\theta z}^{(m)}$ represents the convolution value of the rth convolution kernel of the mth layer to the zth convolution area; S represents the width of the convolution kernel; $C_s^{(m)}(r)$ is the sth weight value of the rth convolution kernel of the mth layer; $p_{zs}^{(m)}(\theta)$ is the θth vector that is convolved in the mth layer, $zs = z * S - S + s$ represents the sth element of the zth convolved part; $p(0)$ is the initial vector.

The rectified linear unit is used as the activation function to perform nonlinear spatial mapping on the convolution result, and the activation result can be expressed as:

$$a_z^{(m)}(r) = \sum_{\theta=1}^{\Theta} \text{relu}\left(y_{\theta z}^{(m)}(r)\right) \tag{2.2}$$

Using the maximum computation with a step size of S1, the activation result is pooled as:

$$p_t^{(m)}(r) = \max_{(t-1)S_1 + 1 < z < tS_1} \left\{ a_z^{(m)}(r) \right\} \tag{2.3}$$

The output of the fully connected layer can be used as the feature extraction result of the entire convolutional neural network, and the vector length can be adjusted and determined according to the actual situation. The one-dimensional convolution kernel has a certain length (sliding window) and a predefined span. The operation process of the vector convolution of the previous layer can be represented in Figure 2.2.

2.2.2 Convolutional Recurrent Neural Network Model

The one-dimensional convolution result is used as the recurrent neural network input, which increases the time-direction analysis while ensuring that the convolution features are effectively used. Therefore, the features of the object are prominent and the diagnostic accuracy is improved. The structure of the convolutional recurrent neural network model is shown in Figure 2.3.

Figure 2.3 shows a convolutional recurrent neural network model with three basic units (the actual number can be different), x is a certain segment of the signal to be analyzed and is fed as input equally to the three convolutional neural network models

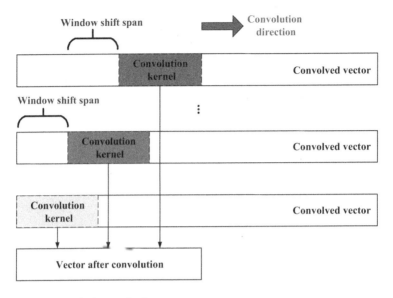

FIGURE 2.2 Convolution method.

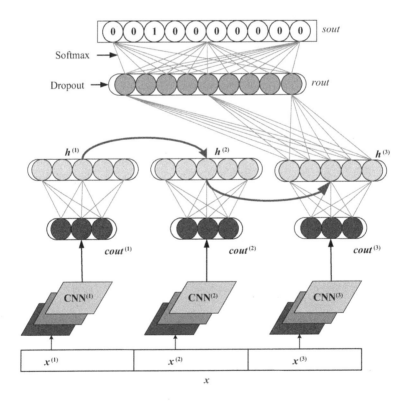

FIGURE 2.3 Structure of convolutional recurrent neural network.

CNN$^{(1)}$ ~ CNN$^{(3)}$ with the same structure. Then the recurrent neural network is used as the receiving end to receive the output of the convolutional neural network to perform in-depth signal analysis on *cout*. Assuming that recurrent neural network has a total of N units, and convolutional neural network has M convolution and pooling layers, the general convolutional recurrent neural network model computation process is given below:

$$\begin{cases} y_{\theta z}^{(n,m)}(r) = \sum_{s=1}^{S} c_s^{(n,m)}(r) p_{zs}^{(n,m-1)}(\theta) \\ a_z^{(n,m)}(r) = \sum_{\theta}^{\Theta} \text{relu}\left(y_{\theta z}^{(n,m)}(r)\right) \\ p_t^{(n,m)}(r) = \max_{(t-1)S_1+1<z<tS_1} \left\{a_z^{(n,m)}(r)\right\} \end{cases} \quad (2.4)$$

where n and m respectively represent the ordinal number of the basic unit of the recurrent neural network and the layer number of the convolutional network, and the equation $p^{(n,0)} = x^{(n)}$ holds.

Expanding the pooling result of the last convolution layer and feeding into the recurrent neural network in vector form as input, the computation method of *rout* can be obtained as follows:

$$\begin{cases} h_i^{(n)} = \tanh\left(\sum_{j=1}^{J} w_{ij} cout_j^{(n)} + \sum_{k=1}^{K} u_{ik} h_k^{(n-1)}\right) \\ rout_l = \text{relu}\left(\sum_{i=1}^{I} v_{li} h_i^{(N)}\right) \end{cases} \quad (2.5)$$

For the classification layer, the softmax function can compress the output of multiple neurons into the (0,1) interval, thereby realizing multi-classification in the form of probability. Assuming that the input of the classification layer is u1 ~uN, the computation process is shown in Figure 2.4.

In theory, softmax is a vector space mapping operation. The main function is to probabilistically concentrate a space vector with multiple parameters into a vector of the same dimension and limit the size of the element to the region (0,1). Softmax can solve the problem of multi-classification and there are no samples failing to be classified as a category. This advantage depends on its special loss function. Different from the "mean square errors" used by conventional neural networks, softmax uses the mean log form of the actually obtained value as the loss function. The expected output *eout* of the softmax layer is a vector represented by a binary number with a

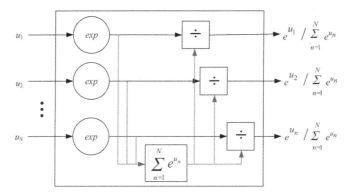

FIGURE 2.4 Structure of softmax.

single bit equal to 1 and the other bits all 0s. The actual output *sout* can be computed using (2.6).

$$\begin{cases} y_g = \sum_{l=1}^{L} \tilde{w}_{gl} rout_l \\ sout_g = y_g \bigg/ \sum_{g=1}^{G} y_g \end{cases} \quad (2.6)$$

The key of network learning is to find the gradient by back propagation and to update and adjust the model parameters according to the gradient direction to find the minimum point of the loss function at the fastest speed, which is called the gradient descent method. Assuming the objective function value is labeled as *loss*, the gradient solution of the *rout* layer can be expressed as:

$$\begin{cases} loss = \frac{1}{G} \sum_{g=1}^{G} \ln(sout_g) \\ \nabla L_{rout} = \frac{dloss}{drout} = sout - eout \end{cases} \quad (2.7)$$

According to the respective characteristics of recurrent neural networks and convolutional neural networks, the proposed convolutional recurrent neural network uses two types of activation functions. The hyperbolic tangent function is derived as $\sigma = \tanh'(u) = 1 - \tanh^2(u)$, and the derivative form of the rectified linear function of the convolutional layer can be expressed as:

$$\text{relu}'(u) = \begin{cases} 0, & u \leq 0 \\ 1, & u > 0 \end{cases} \quad (2.8)$$

To solve the gradient of each layer of neurons and weight variables of the convolutional recurrent neural network, the recurrent neural network must be solved first, and the specific computations are shown in (2.9) ~ (2.11).

$$\begin{cases} \dfrac{\partial rout}{\partial h_i^{(N)}} = \text{relu}'_{rout \to h^{(N)}} \odot V_{\_i} \\[6pt] \nabla L_{h_i^{(N)}} = \nabla L_{rout} \cdot \dfrac{\partial rout}{\partial h_i^{(N)}} \\[6pt] \dfrac{\partial h^{(n)}}{\partial h_i^{(n-1)}} = \sigma_{h^{(n)} \to h^{(n-1)}} \odot U_{\_i} \\[6pt] \nabla L_{h_i^{(n-1)}} = \nabla L_h^{(n)} \cdot \dfrac{\partial h^{(n)}}{\partial h_i^{(n-1)}} \end{cases} \qquad (2.9)$$

where $V_{\_i}$ represents the i-th column of the matrix, and $\odot$ is the symbol of the multiplication of the corresponding position.

$$\begin{cases} \dfrac{\partial rout_l}{\partial v_{li}} = \text{relu}'_{rout_l \to h^{(N)}} \cdot h_i^{(N)} \\[6pt] \nabla L_{v_{li}} = \nabla L_{rout_l} \cdot \dfrac{\partial rout_l}{\partial v_{li}} \\[6pt] \dfrac{\partial h_{i_1}^{(n)}}{\partial u_{i_1 i_2}} = \sigma_{h_{i_1}^{(n)} \to h^{(n-1)}} \cdot h_{i_2}^{(n-1)} \\[6pt] \nabla L_{u_{i_1 i_2}} = \sum_{n=2}^{N} \nabla L_{h_{i_1}}^{(n)} \dfrac{\partial h_{i_1}^{(n)}}{\partial u_{i_1 i_2}} \\[6pt] \dfrac{\partial h_i^{(n)}}{\partial w_{ij}} = \sigma_{h_i^{(n)} \to cout^{(n)}} \cdot cout_j \\[6pt] \nabla L_{w_{ij}} = \sum_{n=1}^{N} \nabla L_{h_i}^{(n)} \dfrac{\partial h_i^{(n)}}{\partial w_{ij}} \end{cases} \qquad (2.10)$$

The gradient of the *cout* layer should be obtained before performing the subsequent computations:

$$\begin{cases} \dfrac{\partial h^{(n)}}{\partial cout_j^{(n)}} = \sigma_{h^{(n)} \to cout^{(n)}} \odot W_{\_j} \\[6pt] \nabla L_{cout_j^{(n)}} = \nabla L_h^{(n)} \cdot \dfrac{\partial h^{(n)}}{\partial cout_j^{(n)}} \end{cases} \qquad (2.11)$$

The gradient of the *cout* layer is used as the loss function of the convolutional neural network layer, which is represented by ∇T here. Next, the gradient of the convolutional neural network can be computed. For the pooling layer, first assume that zt is the position of the maximum value of the t-th pooling area:

$$\nabla T_{a_z^{(n,m)}(r)} = \begin{cases} 0, & z \neq z_t \\ \dfrac{\partial T}{\partial p_t^{(n,m)}}, & z = z_t \end{cases} \tag{2.12}$$

For the convolutional part, the layer element gradient and the convolution kernel gradient can be solved as follows:

$$\begin{cases} \dfrac{\partial a_z^{(n,m)}(r)}{\partial y_{\theta z}^{(n,m)}(r)} = \text{relu}'_{a^{(n,m)} \to y_{\theta z}(r)} \\[6pt] \nabla T_{y_{\theta z}^{(n,m)}(r)} = \nabla T_{a_z^{(n,m)}(r)} \cdot \dfrac{\partial a_z^{(n,m)}(r)}{\partial y_{\theta z}^{(n,m)}(r)} \\[6pt] \nabla T_{p_t^{(n,m-1)}(\theta)} = \sum_{zs=t} C_s^{(m)}(r) \cdot \nabla T_{y_{\theta z}^{(n)}(r)} \\[6pt] \nabla T_{C_s^{(n,m)}(r)} = \sum_{\theta}^{\Theta} \sum_{z}^{Z} \nabla T_{y_z^{(n,m)}(r)} \cdot p_{zs}^{(n,m-1)}(\theta) \end{cases} \tag{2.13}$$

where $\sum_{zs=t}$ represents the sum of all elements satisfying $zs=t$.

2.2.3 DROPOUT IN NEURAL NETWORK MODEL

In the training process of the deep learning model, if the obtained model only fits the training samples well but cannot meet the test requirements of other data, then this phenomenon is called overfitting. Overfitting is usually caused by the limited number of training samples. In recent years, dropout has been widely used and is an effective method among many solutions to eliminate overfitting. Figure 2.5 shows how dropout eliminates overfitting.

Dropout prevents the co-adaptation of neurons by training multiple sub-networks and averaging the results to the entire network set. For the multiple sub-networks as shown in Figure 2.4, some neurons are discarded with probability p; that is, the output value is set to 0 and the remaining neurons are retained with probability $q = 1 - p$.

Assuming that the network input vector is x, the hidden layer can be expressed as $h(x) = f(wx + b)$ in the case of a normal fully connected network. During the dropout training process, there is a random selection of neurons. This method of selection and discarding reduces the complexity of the model while also retaining the information of the input samples, as shown in (2.14).

$$\begin{cases} h(x) = E \odot f(wx + b) \\ E = (e_1, e_2, \ldots, e_i) \end{cases} \tag{2.14}$$

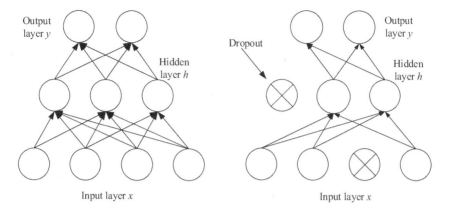

FIGURE 2.5 Dropout to eliminate overfitting.

where E is the i-dimensional Bernoulli vector, the probability that the variable $b = 1$ is p, and the probability that the variable $b = 0$ is $1 - p$.

Using dropout on a certain neuron, the output expression is:

$$o_i = e_i f\left(\sum_{j=1}^{J} w_j x_j + b_j\right) = \begin{cases} f\left(\sum_{j=1}^{J} w_j x_j + b_j\right), & e_i = 1 \\ 0, & e_i = 0 \end{cases} \quad (2.15)$$

Since each neuron maintains the probability of q during the training process, the activation function is appropriately scaled down by the coefficient q in the test phase:

$$o_i = q \; f\left(\sum_{j=1}^{J} w_j x_j + b_j\right), \quad e_i = 1 \quad (2.16)$$

2.3 DIAGNOSTIC FLOW OF THE PROPOSED METHOD

The key to fault diagnosis of rotating machinery based on convolutional recurrent neural network is to be able to use prior data to train a deep learning network that accurately summarizes the working characteristics of machinery. The network can assign a corresponding predefined sample label to new similar data fragments so that the model can be used to identify the types of future data to realize fault detection and fault diagnosis. Taking into account the complex and changeable characteristics of the working environment of rotating machinery, the data under multiple working conditions are collected as input values during training and divided into two parts: a training set and a test set. The entire network training process and fault diagnosis process are shown in Figure 2.6.

In Figure 2.6, the training of the model is completed in the offline stage. For data-driven fault diagnosis methods, obtaining a diagnostic model with high test accuracy

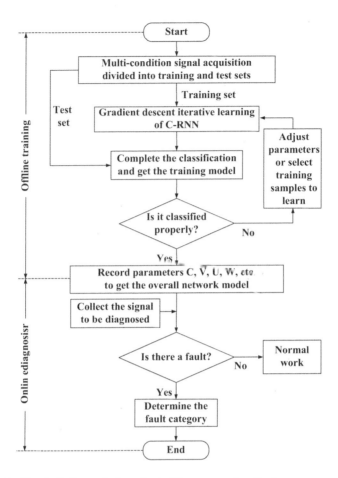

FIGURE 2.6 The fault diagnosis flow chart of the proposed method.

is a key issue and main part of the overall system establishment process. Therefore, the proper signal acquisition operation and parameters tuning process are necessary.

2.4 EXPERIMENTAL RESEARCH BASED ON THE PROPOSED METHOD

2.4.1 Experiment Platform

This experiment uses the Case Western Reserve University data set. The structure of the experiment platform includes power equipment such as drive motors and power meters (to measure the load). The study in this chapter mainly uses the data set of the drive end bearing of the platform at different working speeds and loads. Among them, the fault on the rolling bearing is introduced by an electric spark, and the form of the fault is the damage of different scales on the inner ring, outer ring, and balls. The data set gives enough data on the type and length of the fault.

2.4.2 EXPERIMENTAL SETUP

In order to cope with the problem of the changing working environment of the bearing, the study in this chapter uses the data under four working conditions to train the proposed convolutional recurrent neural network model. The aim is to test whether the fault diagnosis system established in this chapter can still distinguish the bearing category under variable working conditions. The motor power of the platform is 2 hp, and the velocities of the motor are different under the load of 0–3 hp, as shown in Table 2.1.

The working signal of the motor bearing collected by the accelerometer at the motor drive end is selected as the training sample. The sampling frequency is 12 kHz; that is, 12,000 amplitude data can be obtained per second. In this experiment, 2048 data collected within 0.2 seconds are used as a sample to be analyzed, and the sample and label settings for nine types of fault and normal data are shown in Table 2.2.

In Table 2.2, there are 452 samples under each type of label, where 416 samples (104 training samples for each of 4 working conditions) are selected for training and 36 samples (9 test samples for each of 4 working conditions) are selected for testing. The working signal waveforms of various types of bearings are similar. If multiple operating conditions and multiple types of fault scales are considered, it is not easy to intuitively determine the fault type from the signal waveform.

TABLE 2.1
Setting of Operating Conditions

Load (hp)	Velocity (rpm)	Working Condition
0	1797	A
1	1772	B
2	1750	C
3	1730	D

TABLE 2.2
Fault Samples and Label Settings

Label	Fault Location	Size (inch)	Number of Samples
1	Ball	0.007	452
2	Ball	0.014	452
3	Ball	0.021	452
4	Inner race	0.007	452
5	Inner race	0.014	452
6	Inner race	0.021	452
7	Outer race	0.007	452
8	Outer race	0.014	452
9	Outer race	0.021	452
10	Normal	0	452

2.4.3 Summary of Experimental Results

A good convolution recurrent neural network fault diagnosis model can be obtained through experiment tests and parameter tuning and adjustment. The substructure of the proposed method, the recurrent neural network in this experiment has a three-layer network structure of input, hidden, and output layers respectively. The number of neurons in each layer is 108, 278, 100 respectively, and this basic structure was reused four times in the experiment. Another sub-structure of the proposed method, the convolutional neural network has four layers of convolution and pooling operations, and the feature extraction results of the last layer are rearranged into high-dimensional vectors as the input of the recurrent neural network. The specific parameter settings of the convolutional neural network are shown in Table 2.3.

The vector *rout* in Figure 2.3 can be regarded as the feature expression result of the entire convolutional recurrent neural network. The feature vectors with obvious differences are essential for the effective identification of bearing faults. In order to form a comparison, the proposed method in this chapter simultaneously performs visualization analysis on the features of the layer before the classification layer of the conventional convolutional neural and recurrent neural networks and the proposed convolutional recurrent neural network. The visualization analysis adopts the t-distributed stochastic neighbor embedding algorithm, which can map objects with high similarity in the high-dimension space to similar locations in the low-dimensional space with high probability, while keeping objects with low similarity in the high-dimension space stay far away from each other in the low-dimensional space. Considering both accuracy and training time, a relatively good and balanced result can be achieved when the number of convolution layers is eight, and the design of more than eight layers of convolution kernels can not only improve the accuracy but also increase the training time significantly.

As the training target of the network, the loss function is used to evaluate the performance of the network. The closer the value of the loss function is to zero, the better the network performance is to find the most suitable model. In the experiment, the loss functions of all three networks can approach zero after several iterations.

The accuracy and the network depth of the bearing fault identification of the three networks are summarized and compared in Table 2.4. From Table 2.4, all the three network models can obtain high test accuracy. The accuracy of the proposed method

TABLE 2.3
Convolutional Neural Network Settings in the Proposed Model

Layers of Convolutional Neural Network	Convolution Kernel Length	Convolution Kernel Number	Convolution Translation Span	Pooling Width
1	16	16	1	2
2	10	20	1	2
3	8	24	1	2
4	6	18	1	2

TABLE 2.4
Comparison of Deep Learning Methods

Methods	Network Depth	Accuracy
Convolutional neural network	8	94.17%
Recurrent neural network	32	96.67%
Convolutional recurrent neural network	4	99.44%

TABLE 2.5
The Gearbox Fault Data Set

Gear Type	Number of Training Samples	Number of Test Samples	Category Label
Normal	72	24	10000
Tooth missing	72	24	01000
Tooth broken	72	24	00100
Tooth crack	72	24	00010
Tooth worn	72	24	00001

is close to 100.0% (99.44%), which is the highest; and the other two methods can also obtain an accuracy of approximately 95.00%, indicating that the deep learning methods have strong signal analysis and data processing capabilities. As the number of iterative learning increases during offline training, the corresponding test accuracy improves for all three methods. The experiment results illustrate that all the three network models can achieve high accuracy with acceptable training time. The highest diagnostic accuracy of the proposed method shows that the proposed method has strong feature extraction ability and a simple diagnosis process.

Besides the accuracy, the convolution neural network model has eight layers, the recurrent neural network model has 32 basic units, but the proposed convolutional recurrent neural network method only needs a simple structure of four layers. Therefore, the proposed method can achieve significantly better results using a lower network depth compared with the other two methods as shown in Table 2.4.

The proposed convolutional recurrent neural network model has universal applicability and can also deal with other types of fault diagnosis methods, such as the gearbox fault data set shown in Table 2.5. The proposed method can successfully achieve fault diagnosis on this gearbox data set with an accuracy of 95.00%.

BIBLIOGRAPHY

Abdel-Magied, M., Loparo, K. & Lin, W. Fault detection and diagnosis for rotating machinery: A model-based approach. *Proceedings of the 1998 American Control Conference*, 1998. 3291–3296.

Achille, A. & Soatto, S. 2018. Information dropout: Learning optimal representations through noisy computation. *IEEE Transactions on Pattern Analysis and Machine Intelligence*, 40, 2897–2905.

Al Thobiani, F. & Tinga, T. 2017. An approach to fault diagnosis of rotating machinery using the second-order statistical features of thermal images and simplified fuzzy ARTMAP. *Engineering*, 9, 524–539.

Baydar, N., Chen, Q., Ball, A. & Kruger, U. 2001. Detection of incipient tooth defect in helical gears using multivariate statistics. *Mechanical Systems and Signal Processing*, 15, 303–321.

Chen, Z. & Li, Z. Research on fault diagnosis method of rotating machinery based on deep learning. *2017 Prognostics and System Health Management Conference*, 2017. 1–4.

De Bruin, T., Verbert, K. & Babuška, R. 2016. Railway track circuit fault diagnosis using recurrent neural networks. *IEEE Transactions on Neural Networks and Learning Systems*, 28, 523–533.

El-Thalji, I. & Jantunen, E. 2015. A summary of fault modelling and predictive health monitoring of rolling element bearings. *Mechanical Systems and Signal Processing*, 60, 252–272.

Hyers, R., McGowan, J., Sullivan, K., Manwell, J. & Syrett, B. 2006. Condition monitoring and prognosis of utility scale wind turbines. *Energy materials*, 1, 187–203.

Izzularab, M., Aly, G. & Mansour, D. On-line diagnosis of incipient faults and cellulose degradation based on artificial intelligence methods. *Proceedings of the 2004 IEEE International Conference on Solid Dielectrics*, 2004. 767–770.

Janssens, O., Schulz, R., Slavkovikj, V., Stockman, K., Loccufier, M., Van De Walle, R. & Van Hoecke, S. 2015. Thermal image based fault diagnosis for rotating machinery. *Infrared Physics & Technology*, 73, 78–87.

Jun, Y., Zhensheng, F., Xien, Z. & Pengyuan, L. 2001. Study on missile intelligent fault diagnosis system based on fuzzy NN expert system. *Journal of Systems Engineering and Electronics*, 12, 82–87.

Lee, K. B., Cheon, S. & Kim, C. O. 2017. A convolutional neural network for fault classification and diagnosis in semiconductor manufacturing processes. *IEEE Transactions on Semiconductor Manufacturing*, 30, 135–142.

Lei, Y. 2016. *Intelligent fault diagnosis and remaining useful life prediction of rotating machinery*, Butterworth-Heinemann.

Lei, Y., He, Z. & Zi, Y. 2008. Fault diagnosis based on novel hybrid intelligent model. *Chinese Journal of Mechanical Engineering*, 44, 112–117.

Lei, Y., He, Z., Zi, Y. & Hu, Q. 2007. Fault diagnosis of rotating machinery based on multiple ANFIS combination with GAs. *Mechanical Systems and Signal Processing*, 21, 2280–2294.

Liu, Z., Wang, J., Duan, L., Shi, T. & Fu, Q. Infrared image combined with CNN based fault diagnosis for rotating machinery. *2017 International Conference on Sensing, Diagnostics, Prognostics, and Control*, 2017. 137–142.

Lu, D. & Qiao, W. Adaptive feature extraction and SVM classification for real-time fault diagnosis of drivetrain gearboxes. *2013 IEEE Energy Conversion Congress and Exposition*, 2013. 3934–3940.

Marçal, R. F. M., Hatakeyama, K. & Czelusniak, D. J. 2015 Expert System based on fuzzy rules for monitoring and diagnosis of operation conditions in rotating machines. *Advanced Materials Research*. Trans Tech Publ, 950–960.

Martins, J. F., Pires, V. F. & Pires, A. J. 2007. Unsupervised neural-network-based algorithm for an on-line diagnosis of three-phase induction motor stator fault. *IEEE Transactions on Industrial Electronics*, 54, 259–264.

McDuff, R. J., Simpson, P. K. & Gunning, D. An investigation of neural networks for F-16 fault diagnosis. I. system description. *IEEE Automatic Testing Conference. The Systems Readiness Technology Conference. Automatic Testing in the Next Decade and the 21st Century*. Conference Record., 1989. IEEE, 351–357.

Mitoma, T., Wang, H. & Chen, P. 2008. Fault diagnosis and condition surveillance for plant rotating machinery using partially-linearized neural network. *Computers & Industrial Engineering*, 55, 783–794.

Najafi, A., Iskender, I., Farhadi, P. & Najafi, B. Turn to turn fault diagnosis for induction machines based on wavelet transformation and BP neural network. *International Aegean Conference on Electrical Machines and Power Electronics and Electromotion, Joint Conference*, 2011. IEEE, 294–297.

Park, J., Ha, J. M., Oh, H., Youn, B. D., Choi, J.-H. & Kim, N. H. 2016. Model-based fault diagnosis of a planetary gear: A novel approach using transmission error. *IEEE Transactions on Reliability*, 65, 1830–1841.

Peng, Z., Kessissoglou, N. & Cox, M. 2005. A study of the effect of contaminant particles in lubricants using wear debris and vibration condition monitoring techniques. *Wear*, 258, 1651–1662.

Saddam, B., Aissa, A., Ahmed, B. S. & Abdellatif, S. Detection of rotor faults based on Hilbert Transform and neural network for an induction machine. *2017 5th International Conference on Electrical Engineering-Boumerdes (ICEE-B)*, 2017. IEEE, 1–6.

Shaw, D., Al-Khalili, D. & Rozon, C. Accurate CMOS bridge fault modeling with neural network-based VHDL saboteurs. *IEEE/ACM International Conference on Computer Aided Design. IEEE/ACM Digest of Technical Papers*, 2001. IEEE, 531–536.

Smith, J. G., Kamat, S. & Madhavan, K. Model-based Fault Diagnosis of Nonlinear System using Intermediate ANN-Hammerstein Approach. *2006 IEEE International Conference on Industrial Technology*, 2006. IEEE, 1355–1360.

Sun, J., Yan, C. & Wen, J. 2017. Intelligent bearing fault diagnosis method combining compressed data acquisition and deep learning. *IEEE Transactions on Instrumentation and Measurement*, 67, 185–195.

Wang, D., Yang, D., Xu, J. & Xu, K. Computational intelligence based machine fault diagnosis. *Proceedings of the IEEE International Conference on Industrial Technology (ICIT'96)*, 1996. IEEE, 465–469.

Watanabe, K. & Toyoda, N. Diagnosis of machine scream-identification of standard number of bearings and diagnosis of the failure modes. *Proceedings of the 1992 International Conference on Industrial Electronics, Control, Instrumentation, and Automation*, 1992. IEEE, 1058–1063.

Yang, B.-S., Lim, D.-S. & Tan, A. C. C. 2005. VIBEX: An expert system for vibration fault diagnosis of rotating machinery using decision tree and decision table. *Expert Systems with Applications*, 28, 735–742.

Yang, Z.-X. & Zhang, P.-B. ELM Meets RAE-ELM: A hybrid intelligent model for multiple fault diagnosis and remaining useful life predication of rotating machinery. *2016 International Joint Conference on Neural Networks (IJCNN)*, 2016. 2321–2328.

Zhang, X., Luo, J., Jiao, J. & Guan, B. The Application of BP Neural Network in the Fault Diagnosis of Rotating Machinery. *2011 Second International Conference on Digital Manufacturing & Automation*, 2011. IEEE, 1199–1202.

Zhang, Y., Xia, J. & Li, L. Fault diagnosis of hydraulic system based on improved BP neural network technology. *2013 International Conference on Measurement, Information and Control (ICMIC)*, 2013.

Zhao, H., Sun, S. & Jin, B. 2018. Sequential fault diagnosis based on LSTM neural network. *IEEE Access*, 6, 12929–12939.

3 Fault Diagnosis of Rotating Machinery Gear Based on Random Forest Algorithm

3.1 INTRODUCTION

In recent years, the fault diagnosis research for large rotating machinery has attracted more and more people's attention, and many remarkable results have been achieved. As one of the most important components of rotating machinery, gears are likely to cause serious economic losses and safety accidents once they fail. Therefore, it is of great significance to detect potential faults of gears as soon as possible and take corresponding diagnostic measures. However, most of the research works on fault diagnosis of rotating machinery gearbox are based on a single classifier, which may reduce the accuracy of fault diagnosis to a certain extent. The random forest algorithm is an ensemble learning algorithm composed of multiple weak classifiers, and the combination of classifiers can effectively improve the accuracy of classification. In view of this, this chapter applies the random forest algorithm to the fault diagnosis of the rotating machinery gearbox, uses the idea of ensemble learning to improve the prediction accuracy of the model, and conducts experimental verification on the rotating machinery experiment platform using a wind turbine drivetrain diagnostic simulator.

In this proposed method, the sensor signals of the gearbox with multiple operating conditions and multiple faults are collected and the time-domain features are extracted first. Then the data after the feature extraction is used as the input feature of the random forest to construct the ensemble learning fault diagnosis model. Finally, the constructed random forest model is used in the fault diagnosis of the rotating machinery gearbox, and the experimental results are compared with the conventional machine learning method. The results show that the random forest algorithm can diagnose gearbox faults well and avoid the complicated process of finding parameters. The proposed method can handle large-scale data sets through the combination of classifiers and improve the prediction accuracy of the model by avoiding the overfitting phenomenon of the conventional classifier. For new faults in the gearbox of the system or faults in other parts of the rotating machinery, the general fault diagnosis procedure is to extract the time domain features of the faulty data, use the features as the input of the random forest algorithm to reconstruct and

DOI: 10.1201/9781003240754-3

retrain the fault diagnosis model, and use the improved random forest model to identify the fault categories.

3.2 FAULT DIAGNOSIS OF ROTATING MACHINERY GEAR BASED ON RANDOM FOREST ALGORITHM

The general procedure in fault diagnosis of rotating machinery gearbox is divided into the following parts: fault data collection, data analysis and processing, diagnosis model establishment, and online fault diagnosis. In this chapter, the random forest algorithm is used to diagnose the faults of the gears of rotating machinery. Compared with the random forest algorithm, the neural network-based method has limitations such as limited association ability, slow convergence speed in diagnosis and overfitting problems, which may cause low diagnosis accuracy. The support vector machine-based fault diagnosis method adopts the principle of structural risk minimization. The diagnosis outcome is good when the training data set is small, but the performance is inferior when facing a large amount of input data compared with other machine learning methods. Therefore, the random forest algorithm used in this chapter not only solves the problem of slow convergence and overfitting of the neural network-based method during the diagnosis process but also deals with the poor diagnosis effect problem of the support vector machine method with a large amount of input data. The specific steps of fault diagnosis for rotating machinery gearbox based on random forest algorithm are described below:

(1) Acquisition of original data set D: first, select three types of gears with faults: root crack, missing tooth, and chipped tooth for data collection. For each type of fault, select l samples data from the vibration sensor measurement of the rotating machinery and u samples from the torque sensor measurement and record the corresponding fault labels in each sample $y_i(i = 1, 2, \cdots, 3l + 3u)$:

$$y_i \in \{y1, y2, y3\}, i = 1, 2, \cdots, 3l + 3u \tag{3.1}$$

$$D = \{D_1 \cup D_2\} \tag{3.2}$$

$$D_1 = \{(x_i, y_i)\}, i = 1, 2, \cdots, 3l \tag{3.3}$$

$$D_2 = \{(x_i', y_i)\}, i = 3l + 1, 3l + 2, \cdots, 3l + 3u \tag{3.4}$$

where y_1, y_2, and y_3 respectively represent three types of faults: root cracks, missing tooth, and chipped tooth, D_1 is the measurement output of the vibration sensor and D_2 is the measurement output of the torque sensor.

(2) Feature extraction and feature subset selection: due to the characteristics of the rotating machinery fault detection system, the following time-domain

TABLE 3.1
Description of Time Domain Features

Serial Number	Time Domain Metric	Math Expression	Symbol
1	Max	$T_1 = \max(x_i)$	x1
2	Root mean square	$T_2 = \sqrt{\sum_{i=1}^{n}(x_i)^2 / n}$	x2
3	Square root amplitude	$T_3 = \left(\sum_{i=1}^{n}\sqrt{|x_i|}/n\right)^2$	x3
4	Standard deviation	$T_4 = \sqrt{\sum_{i=1}^{n}(x_i - \bar{x})^2 / n}$	x4
5	Peak index	$T_5 = \dfrac{T_1}{T_2}$	x5
6	Margin index	$T_6 = \dfrac{T_1}{T_3}$	x6
7	Absolute mean	$T_7 = \dfrac{1}{n}\sum_{i=1}^{n}|x_i|$	x7

features in Table 3.1 are selected as the splitting attributes in the growth process of the random forest. Then randomly sample k times from the feature set with m features each time and combine them into a feature subset.

(3) Bootstrap resampling and the division of training set and test set: apply the bootstrap resampling method by sampling the original data set k times with replacement, and the obtained data is divided into a training data set and a test data set.

(4) Growth of classification and regression trees and results classification: according to the construction method of classification and regression trees, the optimal split feature t_i and the optimal split value a is selected to split the node. The growth of the classification and regression tree is completed when all m features are used as split nodes. By integrating the prediction results of all leaf nodes of the tree, the classification result of the specific tree can be obtained according to the majority vote results.

(5) Classification results of the random forest: take the constructed k trees as the base classifier group of the random forest and divide the feature data into fault categories. By integration of the majority classification results of the k trees, the random forest classification results and the fault diagnosis outcome can be obtained.

The parameters for a fault diagnosis of a rotating machinery gearbox are described in Table 3.2 and the fault diagnosis procedure is described in Table 3.3.

TABLE 3.2
Parameters of Random Forest-Based Fault Diagnosis

Parameters of Random Forest	Metrics
x_i	The i-th measurement of the vibration sensor
x_i'	The i-th measurement of the torque sensor
y_i	Category label of the i-th data
Oob	Out-of-bag error
K	The optimal number of decision trees
M	The optimal number of eigenvectors

TABLE 3.3
Fault Diagnosis of Rotating Machinery Gear Based On Random Forest Algorithm

Offline Process	
Pre-acquisition process:	1. Data collection: vibration sensor measurement and fault type D_1 and torque sensor measurement and fault type D_2; 2. Feature extraction and feature subset selection; 3. Bootstrap resampling: the bootstrap resampling method is used to repeat the sampling k times, and the bootstrap subset generated by each sampling is divided into the training set and the test set. If the error of the random forest remains unchanged, the sampling is stopped, and the k value at this time is the number of optimal trees; otherwise the sampling continues.

Training process:	1. Input: training data set;
	2. Process:
	  (1) Use the k bootstrap subsets obtained by the bootstrap resampling method to grow k trees to obtain the classification result of each tree;
	  (2) Synthesize the classification results of k trees, and use the majority vote as the classification result of the random forest.
	3. Output: the fault type of the training data set.

Online process	

Testing process:	1. Input: test data set;
	2. Process:
	  (1) Use the test data set to grow each tree and obtain the classification result of each decision tree;
	  (2) Get the classification results of the random forest based on the majority vote on the classification results of each tree.
	3. Output: the fault type of the test data set

3.3 EXPERIMENTAL VERIFICATION

The purpose of the experiment is to identify and classify the fault types of a rotating machinery gearbox. Therefore, three types of faults: root crack, missing tooth, and chipped tooth are selected in the experiment. The experimental data is obtained by two sensors: a vibration sensor and a torque sensor and the experiment details are explained as follows.

3.3.1 Experiment Platform

The experimental platform is the wind turbine drivetrain diagnostic simulator produced by SpectraQuest. The transmission system consists of a planetary gearbox, a two-stage parallel shaft gearbox, a bearing loader, and a single-phase motor. The simulator has different mechanical structure configurations to realize multiple working conditions by regulating the programmable magnetic brake and the single-phase motor. The experimental platform can generate different faulty scenarios.

This experiment is to classify the fault types of the gearbox. Therefore, in the experiment, the gap between gears is set to an appropriate size and the possible faults of other parts are not considered. The data collected by the sensors are fed into the computer through a data acquisition device, which are the measurement outputs of the vibration sensor and the torque sensor respectively. The structure of the experimental platform is shown in Figure 3.1, where 1 is a single-phase motor, 2 is a two-stage parallel shaft gearbox, 3 is a planetary gearbox, 4 is a programmable magnetic brake, and 5 indicates the sensors (torque sensor and vibration sensor from left to right). The platform can detect the operation status of the gearbox under multiple operating conditions and multiple faults. The different types of faulty gears are shown in Figure 3.2, where 1 is a normal gear, 2 is a gear with missing tooth, 3 is a gear with root crack, and 4 is a gear with a chipped tooth.

With the aid of the wind turbine drivetrain diagnostic simulator, four working conditions with combinations of the rotation frequency of 6 Hz and 10 Hz and load voltage of 8 V and 5 V are selected and the data from three faulty types are collected. The data collection process is shown below:

Step 1: Set the gear gap to an appropriate size, install the gear with the corresponding type, and adjust the motor rotation frequency and load voltage to the required values.

FIGURE 3.1 Structure of the experimental platform.

FIGURE 3.2 Different types of gears.

Step 2: Collect the data from the vibration sensor and the torque sensor. In this experiment, the number of samples collected from each sensor under each working condition is 2000.

Step 3: Keep the gear type unchanged but adjust the motor rotation frequency and load voltage to the required working condition and repeat the operation of step 2 until the data collection of the four working conditions is completed.

Step 4: Replace the gear with the required type and repeat steps 2 and 3 to collect the data respectively.

Step 5: According to the working conditions, divide the data collected under the above four working conditions into four groups. Each group contains the data measured by two sensors with three types of gears.

Step 6: According to the time-domain features shown in Table 3.1, extract the fault features from the data measured by the sensors under the four working conditions. Using the above method, a multi-dimensional feature vector can be constructed for the measurement output of each sensor. In order to reduce the dimensionality of the data after feature extraction, every four data samples are processed together in the feature extraction. The extraction process is as follows: according to the time domain features described in Table 3.1, the measurement outputs of vibration sensor and torque sensor are respectively extracted for multi-dimensional feature vectors, denoted as set T. This set T consists of the following seven variables: maximum, root mean square, square root amplitude, standard deviation, peak index, margin index, and absolute average. Different variables represent different physical meanings. Among them, the maximum and absolute mean reflect the maximum and average values of the signals obtained in a period of time, which illustrate the severity of the vibration of the rotating machinery; the root mean square can effectively measure the noise level of the system; the square root amplitude and standard deviation reflect the magnitude and fluctuation of the signal energy in the time domain; the peak index and margin index describe the size and amplitude distribution of the time-domain signal impact capability. All the above variables can be used as evaluation indicators in fault diagnosis.

TABLE 3.4
Division of Training Sets and Testing Sets Under One Working Condition

Gear Type	Number of Training Samples	Number of Training Samples	Label
Root crack	339	161	y1
Missing tooth	342	158	y2
Chipped tooth	348	152	y3

TABLE 3.5
Experimental Data

Serial Number	Rotation Frequency (Hz)	Load Voltage (V)	Fault Type	Symbol
1	6	8	{root crack, missing tooth, chipped tooth}	{y1, y2, y3}
2		5		
3	10	8		
4		5		

Step 7: According to the bootstrap resampling method, the data set after the feature extraction under each working condition is repeatedly sampled to generate k bootstrap subsets, and the training set and the test set are divided according to working conditions. The division of training set and test set under the first working condition is shown in Table 3.4. The experimental data are shown in Table 3.5.

3.3.2 Experimental Results

Based on the steps above, the implementation process of fault diagnosis of rotating machinery gearbox based on random forest algorithm is relatively straightforward. The whole experimental process involves the selection of two parameters: the optimal tree number k and the optimal feature subset m. Among them, the value of optimal tree number k is determined according to the out-of-the-bag error during the experiment. For working condition 1, the time domain feature is extracted according to the previous steps, and the extracted feature vector and category label are used as the input of the random forest. As the number of trees increases, the error rate of the model constantly keeps on increasing and the overall trend keeps decreasing until it stabilizes. At this time, the prediction accuracy of the model remains unchanged. Therefore, the k value at this time is selected as the number of optimal trees in the random forest model, which is $k = 320$ in this experiment. Then the size of the optimal feature subset m is determined as $m = 4$. Then the random forest is constructed using the training data set and verified using the testing data set, and the fault diagnosis results are shown in Tables 3.6 and 3.7 respectively.

TABLE 3.6
Confusion Matrix of Training Data Sets Under Working Condition 1

Actual	Predicted			
	y1	y2	y3	Error rate
y1	334	5	0	0.015
y2	1	341	0	0.003
y3	0	0	348	0

TABLE 3.7
Confusion Matrix of Testing Data Sets Under Working Condition 1

Actual	Predicted			
	y1	y2	y3	Error rate
y1	159	2	0	0.012
y2	2	155	1	0.019
y3	0	0	152	0

It can be seen from the confusion matrix in Table 3.7 that when the random forest is used for fault diagnosis, there are two y1 samples misclassified as y2 but 0 samples misclassified as y3, with the overall error rate of 1.2%; there are two y2 samples misclassified as y1 and one y2 sample misclassified as y3, with the overall error rate of 1.9%. The classification results of y3 are all correct. In general, the prediction accuracy rate in the experiment for working condition 1 is 98.84% using the random forest for fault diagnosis. For the other three working conditions, Tables 3.8–3.10 show the test results of the random forest model.

In the above experiment, the overall fault diagnosis accuracy of all working conditions using random forest is relatively high. It can be seen from the confusion matrix that the random forest has the best fault diagnosis effect on y3, and the diagnosis accuracies of y1 and y2 are similar. Besides the high accuracy, the implementation process of the algorithm is relatively simple as the random forest algorithm only

TABLE 3.8
Confusion Matrix of Testing Data Set Under Working Condition 2

Actual	Predicted			
	y1	y2	y3	Error rate
y1	159	3	0	0.019
y2	0	152	0	0
y3	0	0	135	0

Fault Diagnosis of Rotating Machinery Gear

TABLE 3.9
Confusion Matrix of Testing Data Set Under Working Condition 3

Actual	Predicted			Error rate
	y1	y2	y3	
y1	144	3	0	0.020
y2	6	148	0	0.039
y3	0	0	151	0

TABLE 3.10
Confusion Matrix of Testing Data Set Under Working Condition 4

Actual	Predicted			Error rate
	y1	y2	y3	
y1	144	16	0	0.100
y2	15	132	0	0.102
y3	0	0	131	0

involves the selection of two parameters (the optimal tree number k and the optimal feature subset m) in the training stage.

3.3.3 COMPARISON STUDY

In order to verify that the random forest algorithm can effectively improve the fault diagnosis accuracy in this chapter, comparison experiments are conducted using the support vector machine method to extract the same feature vectors under the same working conditions. Support vector machine is a machine learning method based on the principle of structural risk minimization. The core idea is to find a certain non-linear mapping relationship to map the input vector to a higher-dimensional feature space and find the optimal classification hyperplane in this space. This method is usually used for binary classification problems. Suppose the training set is $\{x_i, y_i\}$, $i = 1, 2, \ldots, n$, where x_i is the i-th sample, $y_i \in \{-1, 1\}$ is the category label of the i-th sample, and n is the number of samples. The samples can be divided into two categories by looking for a hyperplane $W \cdot x + b = 0$ that satisfies (3.5).

$$\begin{cases} W \cdot x_i + b \geq +1, & y_i = +1 \\ W \cdot x_i + b \leq -1, & y_i = -1 \end{cases} \quad (3.5)$$

where W is the normal vector of the hyperplane, and b is the bias.

When the support vector machine solves the multi-classification problem, it realizes the classification of multiple faults by combining multiple binary support vector machines, as shown in Figure 3.3. For the three types of faults: root cracks, missing

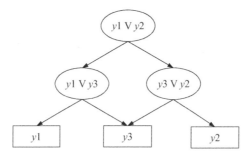

FIGURE 3.3 Multi-classification support vector machine model.

TABLE 3.11
Accuracy Comparison between Random Forest and Support Vector Machine

Serial Number	Rotation Frequency(Hz)	Load Voltage (V)	Random Forest (%)	Support Vector Machine (%)
1	6	8	98.84	92.00
2		5	99.33	91.00
3	10	8	98.01	93.33
4		5	92.92	81.33

tooth, and chipped tooth, the fault data after feature extraction are fed into the model to divide the data into the first and second types according to Figure 3.3. Then the binary classification procedures continue until all the three binary support vector machine models are completed. The following Table 3.11 illustrates the comparison result of the fault diagnosis experiments using the random forest algorithm and the support vector machine method.

It can be seen from the above comparative experiment that under the same working conditions, the fault diagnosis accuracy using the random forest algorithm is significantly higher than that of using the support vector machine method. This is because of the following reasons: the random forest algorithm uses the bootstrap resampling method to generate multiple classifiers, so the training sets of each classifier are different from each other; the feature set is also sampled by random subspace method, so the feature sets of each classifier are also different from each other; the final classification results of the entire forest are also ensemble from the classification results of each classifier. Besides the benefit in accuracy, the time complexity of random forest is also low because the establishment of the random forest model only involves the determination of two parameters. Therefore, compared with the support vector machine method, the fault diagnosis performance of the random forest algorithm is significantly improved.

BIBLIOGRAPHY

Breiman, L. 1996. Bagging predictors. *Machine Learning*, 24, 123–140.
Breiman, L. 2001. Random forests. *Machine Learning*, 45, 5–32.
Chen, X.-W. & Liu, M. 2005. Prediction of protein–protein interactions using random decision forest framework. *Bioinformatics*, 21, 4394–4400.
Fang, K., Wu, J., Zhu, J. & Xie, B. A review of technologies on random forests. Statistics & Information Forum, 2011. 32–38.
Ho, T. K. 1998. The random subspace method for constructing decision forests. *IEEE Transactions on Pattern Analysis and Machine Intelligence*, 20, 832–844.
Lee, S. L. A., Kouzani, A. Z. & Hu, E. J. 2010. Random forest based lung nodule classification aided by clustering. *Computerized Medical Imaging and Graphics*, 34, 535–542.
Lei, Y., He, Z. & Zi, Y. 2008. Fault diagnosis based on novel hybrid intelligent model. *Chinese Journal of Mechanical Engineering*, 44, 112–117.
Li, Y., Cheng, G., Liu, C. & Chen, X. 2018. Study on planetary gear fault diagnosis based on variational mode decomposition and deep neural networks. *Measurement*, 130, 94–104.
Parkhurst, D. F., Brenner, K. P., Dufour, A. P. & Wymer, L. J. 2005. Indicator bacteria at five swimming beaches—analysis using random forests. *Water Research*, 39, 1354–1360.
Strobl, C., Malley, J. & Tutz, G. 2009. An introduction to recursive partitioning: Rationale, application, and characteristics of classification and regression trees, bagging, and random forests. *Psychological Methods*, 14, 323.
Su, T., Ming, Y., Wang, C.-X., Tang, W. & Wang, B. 2018. Classification and regression tree based traffic merging for method self-driving vehicles. *Acta Automatica Sinica*, 44, 35–43.
Xing, Z., Qu, J., Chai, Y., Tang, Q. & Zhou, Y. 2017. Gear fault diagnosis under variable conditions with intrinsic time-scale decomposition-singular value decomposition and support vector machine. *Journal of Mechanical Science and Technology*, 31, 545–553.
Zhi-Ling, Y., Bin, W., Xing-Hui, D. & Hao, L. 2012. Expert system of fault diagnosis for gear box in wind turbine. *Systems Engineering Procedia*, 4, 189–195.
Zhou, D.-H. & Hu, Y.-Y. 2009. Fault diagnosis techniques for dynamic systems. *Acta Automatica Sinica*, 35, 748–758.
Zhuang, J., Luo, J., Peng, Y., Huang, C.-Q. & Wu, C.-Q. 2009. Fault diagnosis method based on modified random forests. *Computer Integrated Manufacturing Systems*, 15, 777–785.

4 Bearing Fault Diagnosis under Different Working Conditions Based on Generative Adversarial Networks

4.1 INTRODUCTION

In the actual operation of rotating machinery, the changes in operating conditions and factors such as load and speed will lead to large distribution differences between the test data set and the training data set, thus making the extracted signals fail to meet the basic assumption in machine learning. The large distribution differences in the two domains can lead to inaccurate fault diagnosis outcome when facing new working conditions.

The existing fault diagnosis for new operating conditions usually requires the labeling of the new operating conditions and recombining the labeled data under the new and existing operating conditions to retrain the previously obtained fault diagnosis model. However, collecting data with tags in new working conditions requires a lot of manpower and material resources in practice, but it is much easier to collect data without tags under new working conditions. Therefore, a fault diagnosis model combining maximum mean discrepancy and Wasserstein generative adversarial network is proposed in this chapter. The model combines domain adaptation and multi-domain adversarial learning and uses labeled data from existing working conditions and unlabeled data from new working conditions to perform fault diagnosis based on transfer learning.

Transfer learning is a type of machine learning and the central idea is to learn relevant knowledge from the source domain and then apply this knowledge to tasks in the target domain to achieve the purpose of improving task performance. In transfer learning, the model transfer and feature transfer approaches are more commonly used than other approaches. The model transfer approach requires sufficient labels in the target domain, but obtaining sample labels requires much manpower and material resources to complete. The feature transfer approach does not require the data of the target domain to have labels and can complete the task through the common features of the source domain and the target domain. In order to achieve this, it is usually in mapping the features of the source domain and the target domain to a certain space that the distribution difference between the source domain features and the target domain features is minimized. The obtained features in this projected space are considered as the common features of the source domain and the target domain which are used in the subsequent learning processes.

Domain adaptation is a typical method of feature transfer approach in transfer learning. The domain adaptation approach aims at the problem that the source domain and the target domain have a common classification task, but the data distribution between the two domains is inconsistent. The idea of domain adaptation is to apply the knowledge of the source domain to the target classification task through the common features of the two domains. The source domain data is often linked with rich tags, but the target domain data has limited or even no tags. At present, the unsupervised domain adaptation for unlabeled target domain data has become a key research issue in fault diagnosis and classification.

4.2 MODEL ESTABLISHMENT AND THEORETICAL DERIVATION

The fault diagnosis model based on the maximum mean discrepancy and Wasserstein generative adversarial network proposed in this chapter is mainly composed of a generative adversarial network guided by the maximum mean difference and Wasserstein distance. The proposed model has unsupervised domain adaptation capability, aiming to solve the problem of bearing fault diagnosis under different working conditions.

4.2.1 Wasserstein Generative Adversarial Network

The generative adversarial network is a powerful generative model in machine learning. The central idea is to use the principle of adversarial learning to make the generated data as similar as possible to real data. The generator network generates synthetic data with some given noise sources, and the discriminator network distinguishes the output of the generator from the real data. In fact, it is the process of a max–min game between the discriminator and the generator, as shown in (4.1):

$$\min_{G} \max_{D} V(D,G) = E_{x \sim p_r}[\log(D(x))] + E_{z \sim p_z}\left[\log(1 - D(G(z)))\right] \qquad (4.1)$$

where $V(D, G)$ is the objective function, D and G are the discriminant model and generative model, p_r and p_z are the distributions of real data and random noise, x and z are the real data and random noise.

The goal of a generative adversarial network is to obtain the optimal Jensen–Shannon divergence between real data and generated data. But because the divergence may be discontinuous with respect to the parameters of the generator, the non-differentiable points may cause the gradient vanishing problem when the discriminator reaches the optimum. Because Wasserstein distance is continuous and differentiable everywhere, it is recommended to use Wasserstein distance, which is also known as the earth mover distance, instead of Jensen–Shannon divergence as a

Generative Adversarial Networks

metric to measure the gap between the real data distribution and the generated data distribution. The computation process is shown in (4.2):

$$W(P_r, P_g) = \inf_{\gamma \in \Pi(P_r, P_g)} E_{(x,y) \in \lambda}\left[\|x - y\|\right] \quad (4.2)$$

where P_r and P_g are the distribution of real sample data and the distribution of generated sample data respectively, x and y are the real samples and generated samples, and $\Pi(P_r, P_g)$ is the set of all samples of the joint distribution $\gamma \in (P_r, P_g)$.

It can be seen from (4.2) that the Wasserstein distance is actually used as a measure to reduce the distance between x and y. Based on the Kantorovich–Rubinstein duality principle, (4.3) can be obtained:

$$W(P_r, P_\theta) = \sup_{\|f\| \leq 1} E_{x \sim P_r}\left[f(x)\right] - E_{x \sim P_\theta}\left[f(x)\right] \quad (4.3)$$

where P_θ is the data distribution of the generated sample, θ is the generator parameter, and $\|f\|$ is the 1-Lipschitz function.

Then the target function in (4.3) can be represented as (4.4):

$$\min_G \max_D E_{x \sim P_r}\left[D(x)\right] - E_{\tilde{x} \sim P_r}\left[D(\tilde{x})\right] \quad (4.4)$$

where D is the set of 1-Lipschitz functions.

The Wasserstein generative adversarial network must tailor the weight of the discriminator to a compact space to implement Lipschitz constraints on the discriminator, but such tailoring will cause optimization difficulty. In order to solve this problem, a gradient penalty coefficient is added to the objective function of the conventional Wasserstein generative adversarial network. Compared with the conventional model, the improvement with gradient penalty coefficient can avoid gradient vanishing and gradient explosion problems, making the network model more robust. The Wasserstein generative adversarial network with gradient penalty is shown in (4.5):

$$\min_G \max_D E_{x \sim P_r}\left[D(x)\right] - E_{\tilde{x} \sim P_r}\left[D(\tilde{x})\right] - \lambda E_{\hat{x} \sim P_{\hat{x}}}\left[\left(\|\nabla_{\hat{x}} D(\tilde{x})\|_2 - 1\right)^2\right] \quad (4.5)$$

where $p_{\hat{x}}$ is the linear uniform sampling between the paired points of the real data distribution and the generated data distribution, and λ is the penalty gradient term.

4.2.2 Maximum Mean Discrepancy

Maximum mean discrepancy is a non-parametric distance metric used to measure the difference in data distributions between two types of data. Assuming the data set $X = \{x_i\}_{i=1}^{m}$ and $Y = \{y_j\}_{j=1}^{n_j}$ satisfying the probability distributions P and Q respectively satisfy the probability distributions P and Q, there is a regenerated kernel Hilbert space H, and there is a mapping $\Phi(\cdot)$ from the original space to the Hilbert space. Let $X \to H$, then the expression of maximum mean discrepancy is shown in (4.6).

$$D_H(X,Y) = \underset{\Phi \in H}{\mathrm{sub}}(E_{X \sim P}\left[\Phi(x)\right] - E_{X \sim Q}[\Phi(y)]) \quad (4.6)$$

where n_1 and n_2 are the number of samples in the data set X and data set Y.

It is noted that the smaller the maximum mean discrepancy is, the smaller the difference in data distribution between the two data sets is. In order to facilitate the computation and application, it is usually to compute the empirical estimation of maximum mean discrepancy using the kernel mean embedding method in (4.7):

$$\hat{D}_H^2(X,Y) = \left\| \frac{1}{n_1} \sum_{i=1}^{n_1} \Phi(x_i) - \frac{1}{n_2} \sum_{i=1}^{n_2} \Phi(y_j) \right\|_H^2 \quad (4.7)$$

4.2.3 Establishment of Fault Diagnosis Model

In this section, a transfer learning-based fault diagnosis model is proposed, based on the maximum mean discrepancy and Wasserstein generative adversarial network, to solve the fault diagnosis problem of rolling bearing under different working conditions. The proposed model is mainly composed of feature extraction module, feature transfer module, and classification module. Among them, the feature extraction module is composed of a five-layer convolutional neural network, and its purpose is to extract transferable features; the feature transfer module is composed of a maximum mean discrepancy optimization layer, two fully connected layers, and Wasserstein generative adversarial network, with the main aim to minimize the distribution difference between the source domain data and the target domain data; the classification module is a fully connected multi-layer neural network. The parameters of the proposed fault diagnosis model are shown in Table 4.1, and the specific parameters of the generative adversarial network are shown in Table 4.2.

TABLE 4.1
The Structure of the Proposed Method

Network Layer Type	Kernel Length	Number of Kernels	Stride	Activation Function
Convolutional layer_1	64	16	2	ReLU
Pooling layer_1	2	16	2	—
Convolutional layer_2	32	32	2	ReLU
Pooling layer_2	2	32	2	—
Convolutional layer_3	16	64	2	ReLU
Pooling layer_3	2	64	2	—
Convolutional layer_4	8	64	2	ReLU
Pooling layer_4	2	64	2	—
Fully connected layer_1	512	1	—	Tanh
Fully connected layer_2	10	1	—	Softmax

TABLE 4.2
The Model Parameters of Generative Adversarial Model

Network Layer Type	Kernel Length	Number of Kernels	Stride	Activation Function
Convolutional layer_1	5	64	2	LeakyReLU
Fully connected layer_1	512	32	—	LeakyReLU
Fully connected layer_2	1	1	—	Linear

4.2.4 Fault Diagnosis Procedures of the Proposed Method

The specific steps of the fault diagnosis procedures of the proposed method in this chapter are shown below:

Step 1: Use labeled source domain data to train a convolutional neural network to obtain good features and a well-trained classification model. The training part of the convolutional neural network uses the cross entropy between the source domain data labels and the prediction of the softmax classifier as the classification loss function to update the parameters of the feature mapping θ^M and the parameters θ^C of the classifier. The cross entropy loss function is shown in (4.8):

$$L_C(x^s, y^s) = -\frac{1}{n^s} \sum_{i=1}^{n^s} \sum_{k=1}^{K} l(y_i^s = k) \cdot \log C(M(x_i^s))_k \qquad (4.8)$$

where K is the number of categories and $\log C(M(x_i^s))_k$ is the probability distribution of the prediction.

Step 2: After completing step 1, the source domain data and target domain data are fed into the pre-trained model. By minimizing the maximum mean discrepancy between the source domain features and the target domain features in the fully connected layers 1 and 2, the target of minimizing the probability difference in data distribution between the two domains can be achieved. The objective function is shown in (4.9):

$$\hat{D}_H^2(X,Y) = \left\| \frac{1}{n_1} \sum_{i=1}^{n} \Phi\left(F_j(M(x_i^s))\right) - \frac{1}{n_2} \sum_{i=1}^{n_2} \Phi\left(F_j(M(x_i^t))\right) \right\|_H^2 \qquad (4.9)$$

where F_j is the output of the j-th fully connected layer:

$$F_j = \begin{cases} F_{c_1}, & j = 1; \\ F_{c_j}(F_{j-1}), & j > 1. \end{cases} \qquad (4.10)$$

Step 3: After minimizing the maximum mean discrepancy, the features are fed into the Wasserstein generative adversarial network, which is composed of a convolutional layer and two fully connected layers. The source domain feature is used as the discriminator, and the target domain feature is used as the data generated by the generator. The Wasserstein generative adversarial network model is used to train the two-layer output of fully connected layers 1 and 2 and compute the Wasserstein distance of the source domain output features and the target domain output features by minimizing the loss function to update the generator parameters θ^d to obtain data with the similar distribution. Then the trained model can be used for fault diagnosis with the test data set to verify the effectiveness of the proposed method. The loss function of the Wasserstein generative adversarial network is shown in (4.11):

$$L_w(x^s, x^t) = \frac{1}{n^s} \sum_{i=1}^{n^s} D\big(F_j(M(x_i^s))\big) - \frac{1}{n^t} \sum_{i=1}^{n^t} D\big(F_j(M(x_i^t))\big) \\ - \lambda E_{\tilde{x} \sim p_{\tilde{x}}} \left[\big(\|(D(\tilde{x}))\|_2 - 1\big)^2 \right] \quad (4.11)$$

4.3 EXPERIMENTAL RESULTS

In this section, the data set is from the test bench established by the Bearing Data Center of Case Western Reserve University obtained under different working conditions. The test bench consisted of a motor, a torque sensor, and a dynamometer, and the vibration data was collected by an accelerometer mounted on the motor housing including both normal and faulty data. Three working conditions were considered in the test bench, as listed in Table 4.3.

The selected training data in this experiment is the vibration signal of the motor drive end, and its sampling frequency is 12 kHz. In this experiment, a training or test sample consists of the data collected by sensors every 0.1 seconds. A total of nine types of faulty and normal conditions are collected, with the total number of samples in each type being 500. The settings of samples and labels are shown in Table 4.4.

In the experiment, the data set in any case from the three working conditions a, b, and c listed in Table 4.3 is used as the source domain D_s, and the other two cases from

TABLE 4.3
Working Conditions of the Test Bench

Work Load (HP)	Speed (rpm)	Working Condition
0	1797	A
1	1772	B
2	1750	C

TABLE 4.4
The Settings of Fault Samples and Labels

Label	Fault Location	Size (inch)	Number of Samples
0	Normal	0	500
1	Rolling element	0.007	500
2	Rolling element	0.014	500
3	Rolling element	0.021	500
4	Inner race	0.007	500
5	Inner race	0.014	500
6	Inner race	0.021	500
7	Outer race	0.007	500
8	Outer race	0.014	500
9	Outer race	0.021	500

TABLE 4.5
Experiment Accuracy

SD ← TD	Accuracy	SD ← TD	Accuracy	SD ← TD	Accuracy
$a \leftarrow b$	99.17%	$b \leftarrow a$	99.48%	$c \leftarrow a$	98.06%
$a \leftarrow c$	99.93%	$b \leftarrow c$	100%	$c \leftarrow b$	97.67%

a, b, and c are used as the target domain D_t, as shown in Table 4.5. The classification accuracy is used as the evaluation metric as shown in (4.12):

$$Acc = \frac{N(x_c)}{N(x)} \qquad (4.12)$$

where $N(x_c)$ is the number of test sets whose predicted label is the same as the actual label, and $N(x)$ is the number of all test data sets.

Table 4.5 shows the accuracy of bearing fault diagnosis in six transfer situations under three working conditions, where SD represents source domain and TD represents target domain. It can be seen from Table 4.5 that the accuracies of most cases are above 99%, and the minimum accuracy can still reach 97.67%. Therefore, the proposed method based on the maximum mean discrepancy and Wasserstein generative adversarial network is suitable for bearing fault diagnosis under different working conditions.

BIBLIOGRAPHY

Ahmed, H., Wong, M. L. D. & Nandi, A. K. 2018. Intelligent condition monitoring method for bearing faults from highly compressed measurements using sparse over-complete features. *Mechanical Systems and Signal Processing*, 99, 459–477.

Arjovsky, M., chintala, S. & Bottou, L. Wasserstein generative adversarial networks. *International Conference on Machine Learning*, 2017. PMLR, 214–223.

Gulrajani, I., Ahmed, F., Arjovsky, M., Dumoulin, V. & Courville, A. 2017. Improved training of wasserstein gans. *arXiv preprint arXiv:1704.00028*.

Guo, L., Lei, Y., Xing, S., Yan, T. & Li, N. 2018. Deep convolutional transfer learning network: A new method for intelligent fault diagnosis of machines with unlabeled data. *IEEE Transactions on Industrial Electronics*, 66, 7316–7325.

Jia, F., Lei, Y., Lu, N. & Xing, S. 2018. Deep normalized convolutional neural network for imbalanced fault classification of machinery and its understanding via visualization. *Mechanical Systems and Signal Processing*, 110, 349–367.

Khan, S. & Yairi, T. 2018. A review on the application of deep learning in system health management. *Mechanical Systems and Signal Processing*, 107, 241–265.

Patel, V. M., Gopalan, R., Li, R. & Chellappa, R. 2015. Visual domain adaptation: A survey of recent advances. *IEEE Signal Processing Magazine*, 32, 53–69.

Sun, S., Zhang, B., Xie, L. & Zhang, Y. 2017. An unsupervised deep domain adaptation approach for robust speech recognition. *Neurocomputing*, 257, 79–87.

Wang, K., Zhao, W., Xu, A, Zeng, P. & Yang, S. 2020. One-dimensional multi-scale domain adaptive network for bearing-fault diagnosis under varying working conditions. *Sensors*, 20, 6039.

Wei, Z., Wang, Y., He, S. & Bao, J. 2017. A novel intelligent method for bearing fault diagnosis based on affinity propagation clustering and adaptive feature selection. *Knowledge-Based Systems*, 116, 1–12.

Wen, L., Gao, L. & Li, X. 2017. A new deep transfer learning based on sparse auto-encoder for fault diagnosis. *IEEE Transactions on Systems, Man, and Cybernetics: Systems*, 49, 136–144.

Yang, B., Lei, Y., Jia, F. & Xing, S. 2019. An intelligent fault diagnosis approach based on transfer learning from laboratory bearings to locomotive bearings. *Mechanical Systems and Signal Processing*, 122, 692–706.

Zhang, B., Li, W., Li, X.-L. & Ng, S.-K. 2018. Intelligent fault diagnosis under varying working conditions based on domain adaptive convolutional neural networks. *IEEE Access*, 6, 66367–66384.

Zhang, L., Zuo, W. & Zhang, D. 2016. LSDT: Latent sparse domain transfer learning for visual adaptation. *IEEE Transactions on Image Processing*, 25, 1177–1191.

Zhao, R., Yan, R., Chen, Z., Mao, K., Wang, P. & Gao, R. X. 2019. Deep learning and its applications to machine health monitoring. *Mechanical Systems and Signal Processing*, 115, 213–237.

Zheng, J., Pan, H. & Cheng, J. 2017. Rolling bearing fault detection and diagnosis based on composite multiscale fuzzy entropy and ensemble support vector machines. *Mechanical Systems and Signal Processing*, 85, 746–759.

5 Rotating Machinery Gearbox Fault Diagnosis Based on One-Dimensional Convolutional Neural Network and Random Forest

5.1 INTRODUCTION

The fault diagnosis of gears is an important part of maintaining the safe operation of rotating machinery. Currently, most of the gear fault diagnosis approaches use the characteristics of vibration signals to classify the gear faults, and most of the existing intelligent fault diagnosis methods use vibration signals to diagnose gear faults, which may not make full use of the information of other sensors. Although some conventional fault diagnosis methods can utilize multi-type sensor signals to diagnose gear faults, these methods require manual feature extraction and selection in fault classification, which require pre-acquired background knowledge and a trivial manual extraction process. With the development of the big data era in the mechanical engineering field, the signals collected by mechanical equipment are no longer limited to vibration signals only but are different types of signals extracted by sensors at different locations, and the overall quantity of these signals is quite large. Therefore, an intelligent fault diagnosis method is needed which can perform intelligent feature extraction and classification at different positions with different types of signals to reduce the manual steps in the corresponding steps due to the development of intelligent methods and the big data era in mechanical engineering.

The convolutional neural network is an intelligent algorithm that has undergone a period of development. Due to the powerful feature expression and feature extraction capabilities, the convolutional neural network is quickly applied to tasks such as image recognition and classification. The local receptive field of the convolutional neural network can automatically extract the features of different regions of the input data and can combine these local features to restore the entire feature region. This ability of convolutional neural networks can not only reduce the number of network parameters but also greatly reduce the risk of network overfitting. Therefore, the convolutional neural networks can be very helpful when used in various application fields due to their powerful feature expression and feature extraction capabilities and have also attracted much attention in the field of fault diagnosis of rotating machinery.

The conventional manual feature extraction methods require a large amount of professional knowledge. In order to reduce the amount of computation, further feature selection is also necessary before the fault classification. Although the convolutional neural networks can achieve end-to-end fault diagnosis and reduce network parameters and the risk of overfitting with the support of local receptive fields, the over-fitting may still occur due to improper selection of network parameters or the number of network layers when applied to the field of fault diagnosis. An intelligent fault diagnosis algorithm, which can automatically realize feature extraction and fault diagnosis without manual feature selection or not heavily relying on network parameters, is required based on the above analysis. Therefore, the random forest algorithm is considered to deal with the above-mentioned problems.

Compared with existing classification algorithms, the random forest algorithm is an ensemble algorithm with high classification accuracy. This algorithm can maintain high-efficiency performance on a large amount of data and is prone to the overfitting problem. This algorithm can also avoid the feature selection step and provide an importance evaluation to each variable during the classification process. Therefore, a fault diagnosis method based on the combination of one-dimensional convolutional neural network and random forest for rotating machinery gearbox fault diagnosis is proposed in this chapter to deal with the above-mentioned problems. The experimental results show that this method can effectively solve the fault diagnosis of rotating machinery with multiple operating conditions, multiple faults, and multiple sensor signals.

5.2 MODEL ESTABLISHMENT AND THEORETICAL DERIVATION

The fault diagnosis method combining one-dimensional convolutional neural network and random forest proposed in this chapter is mainly composed of the feature extraction module of a one-dimensional convolutional neural network and the classification module of random forest. After improvement and combination, the proposed fault diagnosis approach for rotating machinery gearbox has the following advantages:

(1) The proposed model with a one-dimensional convolutional neural network and random forest can directly perform feature extraction and classification of various time-domain signals, avoiding the cumbersome process of manual feature construction and selection process.
(2) The proposed model with a one-dimensional convolutional neural network and random forest has little dependence on network parameters, can reduce the risk of network model overfitting, and can perform efficient and accurate fault diagnosis for rotating machinery gearboxes with multiple operating conditions, multiple faults, and multiple sensor signals.

5.2.1 ONE-DIMENSIONAL CONVOLUTIONAL NEURAL NETWORK

The signals collected in this chapter are stored in one-dimension form and the one-dimensional convolutional neural networks used to process the collected signals are similar to conventional convolutional neural networks, including five parts: input layer, convolution layer, pooling layer, fully connected layer, and output layer. The feature extraction on the original signal is conducted through the convolution operation and the specific operation is shown in (5.1):

$$y_r^{l(i,j)} = \sum_{s=1}^{S} k_r^{l(s)} x^{(l-1)}(i) \quad (5.1)$$

where $y_r^{l(i,j)}$ represents the convolutional output value of the convolution area j of the convolution kernel r of the layer l, $k_r^{l(s)}$ represents the sth weight of the convolution kernel r of the layer l, $x^{(l-1)}(i)$ represents the ith convolved vector of the layer l, and S is the convolution kernel width.

In one-dimensional convolutional neural networks, the rectified linear unit function is selected as the activation function. As some outputs of the rectified linear unit activation function are zero, this can effectively improve network sparsity and reduce overfitting. The expression of the activation function is shown in (5.2):

$$a^{l(i,j)} = \sum_{i}^{l} \text{Relu}\left(y_r^{l(i,j)}\right) = \begin{cases} y_r^{l(i,j)}, & y_r^{l(i,j)} > 0; \\ 0, & y_r^{l(i,j)} \leq 0. \end{cases} \quad (5.2)$$

where $a^{l(i,j)}$ is the activation value of the convolutional layer output $y_r^{l(i,j)}$.

For the pooling layer, the maximum pooling method is adopted which has the advantage of reducing the number of training times to obtain features that are independent of location and improve the robustness of features. The expression of the maximum pooling is shown in (5.3):

$$p^{l(i,j)} = \max_{(j-1)S_1+1 \leq j \leq jS_1} \left\{ a^{l(i,j)} \right\} \quad (5.3)$$

where $p^{l(i,j)}$ represents the output of the pooling area, $a^{l(i,j)}$ is the activation value of layer l, and S_1 is the pooling step.

The fully connected layer is composed of fully connected neural networks, and the function is to expand the pooled output into a one-dimensional vector and classify the input features. For the output layer, the softmax function is adopted as the classification function. The purpose of the softmax function is to convert the neuron input into a probability distribution with the sum of one for the establishment of multi-classification tasks. The computation of the fully connected layer is shown in (5.4):

$$z_j^{l+1} = \sum_{i=1}^{m} W_{ij}^l a_i^l + b_j^l \quad (5.4)$$

where z_j^{l+1} is the output value of the jth neuron of the layer $l + 1$, W_{ij}^l is the weight between the ith neuron and the jth neuron in the layer l and b_j^l is the bias in the lth layer to the jth neuron in the $l + 1$th layer.

As the activation function of the output layer ($l + 1$th layer), the softmax function can be computed using (5.5):

$$q_j = softmax(z_j^{l+1}) = \frac{e^{z_j^{l+1}}}{\sum_k^5 e^{z_k^{l+1}}} \qquad (5.5)$$

In the network training, the objective function is necessary to evaluate whether the output value is consistent with the target value, and this function is also called the loss function. In this chapter, the cross-entropy loss function is adopted as the objective function. For the evaluation criterion, when the output value is consistent with the target value, the output of the loss function is one; otherwise, the output is zero. The expression of the cross-entropy function is shown in (5.6):

$$L = -\frac{1}{n}\sum_{k=1}^{n}\sum_{j} p_k^j \log q_k^j \qquad (5.6)$$

where q_k^j is the actual output value of softmax, p_k^j is the target distribution, and n is the size of the input batch.

5.2.2 Random Forest Algorithm

Random forest algorithm is a machine learning classification method combining bagging algorithm with random subspace algorithm. The algorithm first forms a forest composed of multiple decision trees and then reduces the overfitting problem of a single decision tree by combining the results of multiple decision trees. The steps of the random forest algorithm are listed in Table 5.1.

The classification and regression trees algorithm in Table 5.1 used for feature selection is based on the Gini coefficient. Specifically, assuming there are K categories, p_k is the probability of the k-th category, and the computation of Gini coefficient is shown in (5.7):

$$Gini = 1 - \sum_{k=1}^{K} p_k^2 \qquad (5.7)$$

For the original data set D with total data number $|D|$, assuming that D is divided into K categories, and the number of the k-th category is $|C_k|$, the Gini coefficient of D is shown in formula (5.8):

$$Gini(D) = 1 - \sum_{k=1}^{K} \left(\frac{|C_k|}{|D|}\right)^2 \qquad (5.8)$$

TABLE 5.1
Flow of Random Forest Algorithm

Algorithm of Random Forest

Input:
(1) The original data set D with the number of samples N and the feature attribute F;
(2) The number of models in the combined classifier K;
(3) Decision tree learning algorithm.

Output:
The classification result of the random forest

For 1 to k do
According to the Bootstrap principle, m ($m < M$) subsets are randomly selected from the original data set D with a number of samples of N and feature attribute of F with replacement;
Randomly select f feature attributes to form the training subset as input, and generate K decision trees according to the classification and regression trees algorithm;
End for

Use K decision trees for classification and return the majority vote.

Assuming that the data set D is divided into D_1 and D_2 according to the attribute a of the feature A, the Gini coefficient of D under the condition of A is shown in (5.9):

$$Gini(D,A) = \frac{|D_1|}{|D|}Gini(D_1) + \frac{|D_2|}{|D|}Gini(D_2) \tag{5.9}$$

5.2.3 THE PROPOSED FAULT DIAGNOSIS MODEL

The proposed fault diagnosis model consists of a feature extraction module and a fault classification module. The feature extraction module is composed of five convolutional layers, five pooling layers, one fully connected layer and one output layer. The classification module is a random forest classifier. The operating process of the feature extraction module is to input the diagnostic signal to the convolutional layer, perform batch normalization operation on the convolutional layer output, apply rectified linear unit function for activation and use the maximum pooling operation as down-sampling. This process is repeated until the last pooling layer. Then the output of the last pooling layer is used as the input of the fully connected layer, activated by rectified linear unit, and then fed into the output layer. Finally, the network parameters are updated through backpropagation according to the output results. After multiple iterations, the features extracted are fed into the random forest classifier to obtain the final classification. The specific parameters of the structure of the proposed method are shown in Table 5.2.

5.2.4 ERROR BACK PROPAGATION OF THE PROPOSED MODEL

The error back propagation is the key to the model training, as the update and optimization of model parameters are closely related to the error back propagation

TABLE 5.2
The Specific Parameters of the Structure of the Proposed Method

Layer Number	Network Layer	Number of Convolution Kernels	Convolution Kernel Size	Stride
1	Convolutional layer_1	16	64	8
2	Pooling layer_1	2	2	2
3	Convolutional layer_2	32	5	1
4	Pooling layer_2	2	2	2
5	Convolutional layer_3	32	5	1
6	Pooling layer_3	2	2	2
7	Convolutional layer_4	32	5	1
8	Pooling layer_4	2	2	2
9	Convolutional layer_5	64	5	1
10	Pooling layer_5	2	2	1
11	Fully connected layer	1	100	\
12	Output layer	1	5	\
13	Classification layer	80 (number of decision trees)		

algorithm. Starting from the loss function, the method of error back propagation is to compute the derivative of the weight value and other parameters layer by layer from the backward direction. The computation of model back propagation is shown below.

(1) Back propagation of the fully connected layer

Assume L is the loss function, W_{ij}^l is the weight of the fully connected layer, and b_j^l is the bias of the fully connected layer. Then the derivative of L with respect to these two parameters is calculated in (5.10) and (5.11):

$$\frac{\partial L}{\partial W_{ij}^l} = \frac{\partial L}{\partial z_j^{l+1}} \cdot \frac{\partial z_j^{l+1}}{\partial W_{ij}^l} = \frac{\partial L}{\partial z_j^{l+1}} \cdot a_i^l \tag{5.10}$$

$$\frac{\partial L}{\partial b_j^l} = \frac{\partial L}{\partial z_j^{l+1}} \cdot \frac{\partial z_j^{l+1}}{\partial b_j^l} = \frac{\partial L}{\partial z_j^{l+1}} \tag{5.11}$$

It can be seen from (5.10) and (5.11) that to fully calculate the derivative of L with respect to the two parameters, it is necessary to obtain $\frac{\partial L}{\partial z_j^{l+1}}$, where z_j^{l+1} is the output value of the output layer, and $\frac{\partial L}{\partial z_j^{l+1}}$ is calculated as shown in (5.12):

$$\frac{\partial L}{\partial z_j^{l+1}} = \sum_{k=1}^{n} -\frac{p_k^j}{q_k^j} \left[q_k^j \left(1 - q_k^j\right) \right] = \sum_{k=1}^{n} p_k^j q_k^j - p_k^j \tag{5.12}$$

In addition to the output layer and fully connected layer, the proposed network model also contains a fully connected network with a hidden layer. So it is necessary to calculate the derivative of the loss function L with respect to the weight W_{ij}^{l-1} and bias b_j^{l-1} of the hidden layer. The derivatives of the loss function L with respect to the activation value a_i^l of the hidden layer and the output value z_j^l of that layer are shown in (5.13) and (5.14) respectively:

$$\frac{\partial L}{\partial a_i^l} = \sum_j \frac{\partial L}{\partial z_j^{l+1}} \cdot \frac{\partial z_j^{l+1}}{\partial a_i^l} = \sum_j \frac{\partial L}{\partial z_j^{l+1}} \cdot W_{ij}^l \tag{5.13}$$

$$\frac{\partial L}{\partial z_i^l} = \frac{\partial L}{\partial a_i^l} \cdot \frac{\partial a_i^l}{\partial z_i^l} = \begin{cases} 0, & z_i^l \le 0 \\ \dfrac{\partial L}{\partial a_i^l}, & z_i^l > 0 \end{cases} \tag{5.14}$$

After obtaining $\dfrac{\partial L}{\partial z_i^l}$ according to (5.14), the derivatives of weight W_{ij}^{l-1} and bias b_j^{l-1} can be obtained using (5.10) and (5.11).

(2) Back propagation of the pooling layer

Because the pooling layer does not have parameters such as weights, it is only necessary to calculate the derivative of the loss function L with respect to the neuron input in the back propagation of the pooling layer, as shown in (5.15):

$$\frac{\partial L}{\partial a^{l(i,j)}} = \frac{\partial L}{\partial p^{l(i,j)}} \cdot \frac{\partial p^{l(i,j)}}{\partial^{l(i,j)}} = \begin{cases} 0, & t \ne t_{\max} \\ \dfrac{\partial L}{\partial p^{l(i,j)}}, & t = t_{\max} \end{cases} \tag{5.15}$$

where $t_{\max}$ is the position of the maximum value in this area because the maximum pooling is selected as the pooling function.

(3) Back propagation of the convolutional layer

The parameters of the convolution layer that need to be used in the back propagation are the input and output of the convolution layer and the weight of the convolution kernel. The back propagation computations are shown in (5.16) to (5.18):

$$\frac{\partial L}{\partial y_r^{l(i,j)}} = \frac{\partial L}{\partial a^{l(i,j)}} \cdot \frac{\partial a^{l(i,j)}}{\partial y_r^{l(i,j)}} = \begin{cases} 0, & y_r^{l(i,j)} \le 0 \\ \dfrac{\partial L}{\partial a^{l(i,j)}}, & y_r^{l(i,j)} > 0 \end{cases} \tag{5.16}$$

$$\frac{\partial L}{\partial K_r^{l(s)}} = \frac{\partial L}{\partial y_r^{l(i,j)}} \cdot \frac{\partial y_r^{l(i,j)}}{\partial K_r^{l(s)}} = \sum_i x^{(l-1)}(i) \tag{5.17}$$

$$\frac{\partial L}{\partial x^{(l-1)}(i)} = \sum_i \frac{\partial L}{\partial y_r^{l(i,j)}} \cdot \frac{\partial y_r^{l(i,j)}}{\partial x^{(l-1)}(i)} = \sum_i \frac{\partial L}{\partial y_r^{l(i,j)}} \cdot \sum_{s=1}^{S} K_r^{l(s)} \tag{5.18}$$

5.2.5 Weights Optimization Using Adaptive Moments

To find the optimal weights, the adaptive moments algorithm is adopted as the optimization algorithm in this chapter. This algorithm updates the network parameters by minimizing the loss function and the computation process is shown in (5.19) to (5.22):

$$s_t = \beta_1 s_{t-1} + (1-\beta_1) g_t \tag{5.19}$$

$$v_t = \beta_2 v_{t-1} + (1-\beta_1) g_t^2 \tag{5.20}$$

$$\overline{s_t} = \frac{s_t}{1-\beta_1^t}, \overline{v_t} = \frac{v_t}{1-\beta_2^t} \tag{5.21}$$

$$\theta_{t+1} = \theta_t - \frac{\eta}{\sqrt{v_t + \varepsilon}} \overline{s_t} \tag{5.22}$$

where s_t is the first-order momentum term, v_t is the second-order momentum term, β_1, β_2 are momentum values, $\overline{s_t}, \overline{v_t}$ are first-order and second-order correction values, g_t is the gradient size of the loss function with respect to θ at the t-th iteration, and θ_t is the parameter of the network model at the t-th iteration.

5.3 EXPERIMENTAL RESULTS

5.3.1 Experimental Platform

The experimental platform is the wind turbine drivetrain diagnostic simulator produced by SpectraQuest. The structure is shown in Figure 5.1 and the main structure includes single-phase motors (①), sensors (②), parallel shaft gearbox (③), planetary

FIGURE 5.1 Structure of the experimental platform.

gearbox (④), and variable load brake (⑤). The experimental platform can simulate the situations of multiple operating conditions and multiple gearbox faults of rotating machinery.

5.3.2 Experimental Setup

In reality, the working environment of rotating machinery is complicated. The fault of the rotating machinery may often cause changes in working conditions. Therefore, the fault diagnosis method proposed in this chapter mainly aims at the fault diagnosis of different gear faults under various working conditions. The multiple working conditions can be realized by changing the load and speed of the experimental platform, which can be achieved by changing the input frequency of the motor and the applied voltage of the load. In the setting of the working conditions, six different working conditions (with load voltage 5 V and 8 V and the motor input frequency 6 Hz, 10 Hz, and 14 Hz) are realized through the wind turbine drivetrain diagnostic simulator. The settings of the specific working condition are shown in Table 5.3.

In the gearbox, the driving wheel drives multiple planetary gears to rotate, which has a high faulty probability when the load increases. Therefore, the faults investigated in this chapter are normal, missing tooth, chipped tooth, cracked tooth, and worn tooth of the driving wheel. The data collected by four sensors (two acceleration sensors, one torque sensor, and one pressure sensor) in the wind turbine drivetrain diagnostic simulator are selected as training samples and the sampling frequency is 5120 Hz. In this experiment, the number of samples under each fault type is 720, where there are six working conditions with 120 samples under each working condition, as shown in Table 5.4.

5.3.3 Analysis of Experimental Results

The method based on a one-dimensional convolutional neural network and random forest avoids manual feature extraction and selection. After the feature extraction by a one-dimensional convolutional neural network composed of 5 layers, the extracted features are fed into a random forest classifier composed of 80 decision trees for fault diagnosis. In this chapter, two comparison studies are conducted to show the effectiveness of the proposed method. In comparison study 1, the experimental results on

TABLE 5.3
Setting of Working Conditions

Working Condition Category	Motor Input Frequency (Hz)	Load Voltage (V)
a	6	5
b	6	8
c	10	5
d	10	8
e	14	5
f	14	8

TABLE 5.4
Settings of Samples and Labels

Label	Type of Fault	Number of Samples
1	Normal	720
2	Missing tooth	720
3	Chipped tooth	720
4	Cracked tooth	720
5	Worn tooth	720

TABLE 5.5
Experimental Result of Comparison Study 1

Signal Type	Single Vibration Signal	Combined Signal of Multiple Sensors
Accuracy	89.97%	100%

TABLE 5.6
Experimental Result of Comparison Study 2

Fault Diagnosis Classifier	Test 1	Test 2	Test 3	Average
Softmax	98.91%	98.96%	99.02%	98.96%
Support vector machine	99.20%	99.25%	99.18%	99.21%
Random forest	100%	100%	100%	100%

the single vibration signal and the combined signal using the proposed method are compared; in comparison study 2, the experimental results on different fault diagnosis classifiers (random forest, softmax, and support vector machine) are compared to show the effectiveness of the proposed method.

In comparison study 1, the fault diagnosis method is the same, and the difference is the type of the input data: single vibration signal and combined signal of multiple sensors. The fault diagnosis results of comparison study 1 are shown in Table 5.5. It can be seen from Table 5.5 that the fault diagnosis accuracy with a single vibration signal is only 89.97%, while the combined signal of multiple sensors is 100%. Therefore, compared with a single vibration signal, the fault diagnosis on the combined signal of multiple sensors has better performance.

In order to verify the overall advantages and disadvantages of the proposed fault diagnosis model, the same experimental data is used in comparison study 2 with different methods. The experiment is repeated three times and the results are shown in Table 5.6. In Table 5.6, the average fault diagnosis accuracies based on the softmax classifier, support vector machine classifier, and random forest classifier are 98.96%, 99.21%, and 100% respectively. Therefore, the experimental results show that the method proposed in this chapter can effectively diagnose the faults of rotating machinery gears.

BIBLIOGRAPHY

Breiman, L. 2001. Random forests machine learning. *Journal of Clinical Microbiology*, 45, 5–32.
Guo, L., Gao, H., Zhang, Y. & Huang, H. 2016. Research on bearing condition monitoring based on deep learning. *Journal of Vibration and Shock*, 35, 166–170.
Hao, S., Ge, F.-X., Li, Y. & Jiang, J. 2020. Multisensor bearing fault diagnosis based on one-dimensional convolutional long short-term memory networks. *Measurement*, 159, 107802.
Hu, Q., Si, X.-S., Zhang, Q.-H. & Qin, A.-S. 2020. A rotating machinery fault diagnosis method based on multi-scale dimensionless indicators and random forests. *Mechanical Systems and Signal Processing*, 139, 106609.
Knauer, U., Von Rekowski, C. S., Stecklina, M., Krokotsch, T., Pham Minh, T., Hauffe, V., Kilias, D., Ehrhardt, I., Sagischewski, H. & Chmara, S. 2019. Tree species classification based on hybrid ensembles of a convolutional neural network (CNN) and random forest classifiers. *Remote Sensing*, 11, 2788.
Li, T., Leng, J., Kong, L., Guo, S., Bai, G. & Wang, K. 2019a. DCNR: Deep cube CNN with random forest for hyperspectral image classification. *Multimedia Tools and Applications*, 78, 3411–3433.
Li, Y., Zou, L., Jiang, L. & Zhou, X. 2019b. Fault diagnosis of rotating machinery based on combination of deep belief network and one-dimensional convolutional neural network. *IEEE Access*, 7, 165710–165723.
Park, J., Ha, J. M., Oh, H., Youn, B. D., Choi, J.-H. & Kim, N. H. 2016. Model-based fault diagnosis of a planetary gear: A novel approach using transmission error. *IEEE Transactions on Reliability*, 65, 1830–1841.
Sun, Y., Zhang, H., Zhao, T., Zou, Z., Shen, B. & Yang, L. 2020. A new convolutional neural network with random forest method for hydrogen sensor fault diagnosis. *IEEE Access*, 8, 85421–85430.
Tang, J., Wu, J., Hu, B., Guo, C. & Zhang, J. 2020. A fault diagnosis method using Interval coded deep belief network. *Journal of Mechanical Science & Technology*, 34, 1949–1956.
Xia, M., Li, T., Xu, L., Liu, L. & De Silva, C. W. 2017. Fault diagnosis for rotating machinery using multiple sensors and convolutional neural networks. *IEEE/ASME Transactions on Mechatronics*, 23, 101–110.
Xu, G., Liu, M., Jiang, Z., Söffker, D. & Shen, W. 2019. Bearing fault diagnosis method based on deep convolutional neural network and random forest ensemble learning. *Sensors*, 19, 1088.

6 Fault Diagnosis for Rotating Machinery Gearbox Based on Improved Random Forest Algorithm

6.1 INTRODUCTION

Random forest is a supervised learning algorithm that can handle large-scale data and has good prediction accuracy under data missing situation. However, in the industrial process, the collected samples are usually unlabeled. For these unlabeled samples, people often use semi-supervised learning methods to train classification models with a large number of unlabeled samples and a small number of labeled samples to improve the system performance. At present, the commonly used semi-supervised learning methods mainly include generative model-based method and graph-based method, collaborative training method, etc. Among them, the graph-based semi-supervised learning is one of the most important methods. This graph-based semi-supervised method uses labeled samples and unlabeled samples as the nodes of the graph and assumes similar samples should have the same label. The connection between nodes is built based on the similarity weight matrix between the samples to improve the accuracy of classification. This approach has been widely used in many fields because of its good prediction performance.

As an indispensable transmission and connection part of rotating machinery, the fault and failure of the gearbox can greatly affect the normal operation and the working performance of the entire rotating machinery system. Therefore, the fault diagnosis of the gearbox has been paid close attention by researchers and practitioners in rotating machinery. In the fault diagnosis process of the gearbox, the data must be collected first through the sensors installed during the experiment. Then, the collected data is processed and the fault feature is extracted by the feature extraction method. Finally, a diagnosis model can be established to realize the classification of the fault categories of the collected data.

However, when the conventional random forest algorithm is used for fault diagnosis of the gearbox with unlabeled data, the accuracy of classification cannot be guaranteed. This is because the conventional random forest is a supervised learning algorithm, and the data collected by sensors from the rotational machines are usually unlabeled. It is also very expensive to manually label a large amount of unlabeled data. Motivated by above-mentioned problems, this chapter introduces the idea of semi-supervised learning into the random forest algorithm to solve the problem which the conventional random forest algorithm is difficult to solve and to accurately classify a large amount of unlabeled data.

In order to solve the above-mentioned problems, the following processing steps are required to be done before implementing the algorithm. First, a small part of the collected data is randomly selected for manual labeling. Then, the graph-based semi-supervised learning method is used to predict the labels of the unlabeled samples, and the predicted samples and the original labeled samples are used to build the decision trees. Finally, the prediction results of the decision trees are compared with the semi-supervised prediction results, and the samples with the same predicted labels from both methods are selected and added to the labeled data set to train the random forest together.

Based on the above analysis, this chapter focuses on the problem of low classification accuracy when the conventional random forest algorithm is used for a small number of labeled samples. Semi supervised learning is introduced in the proposed algorithm when applied to the fault diagnosis of rotating machinery. The results show that the improved random forest algorithm can achieve high prediction accuracy of unlabeled sample and have good fault diagnosis performance, which illustrates the important value of the proposed improved random forest algorithm in research.

6.2 IMPROVED RANDOM FOREST ALGORITHM

In this section, the label prediction method with graph-based semi-supervised learning is introduced on the label prediction of unlabeled samples. Then, the semi-supervised learning is combined with the conventional random forest algorithm, and the specific implementation steps of the improved random forest algorithm are described in detail.

6.2.1 Semi-Supervised Learning

Graph-based semi-supervised learning has fast computation speed and high accuracy, and its theory is based on a solid mathematical foundation, which has attracted attention from many researchers in recent years. Graph-based semi-supervised learning is essentially the process of label propagation. First, a graph is established based on the structure between samples, where the nodes of the graph are all sample points and the similarity between sample points is measured by a weight matrix. The commonly used weight matrices are the Gaussian function and Euclidean distance. Then the objective function is established to ensure that similar samples have similar labels and the predicted results on the labeled samples are consistent with the real results. Finally, the function is optimized to obtain the label of the unlabeled samples. The specific implementation process is as follows.

Given a labeled sample set D_l and an unlabeled sample set D_u, where $l + u = m$ and $l \ll u$ as follows:

$$D_l = \{(x_1, y_1), (x_2, y_2), \ldots, (x_l, y_l)\} \quad (6.1)$$

$$D_u = \{x_{l+1}, x_{l+2}, \ldots, x_{l+u}\} \quad (6.2)$$

where $x_i (i = 1, 2, \cdots, m)$ is the given sampling point and $y_i (i = 1, 2, \cdots, l)$ is the category label corresponding to the labeled samples.

Then randomly select a part of the data from the data set D_u to form the data set D_u^1 and construct a graph $G = (V, E)$ based on the data set $D_l \cup D_u^1$, where $V = \{x_1, \ldots, x_l, x_{l+1}, \ldots, x_{l+u}\}$ is the node set, and the edge set E is defined as the following Gaussian function:

$$(W)_{ij} = \begin{cases} \exp\left(\dfrac{-\|x_i - x_j\|_2^2}{2\sigma^2}\right), & \text{if } i \neq j; \\ 0, & \text{otherwise} \end{cases} \quad (6.3)$$

where $i, j \in \{1, 2, \ldots, m\}$, $\sigma > 0$ is the variance of the Gaussian function.

Define a $m \times C$ dimensional label matrix $F = (F_1^T, F_2^T, \cdots, F_m^T)^T$, where C is the number of fault categories and F_i is the label vector of the sample x_i, which is obviously non-negative. The label matrix F can be initialized as follows:

$$F(0) = (Y)_{ij} = \begin{cases} 1, & \text{if } (1 \leq i \leq l) \wedge (y_i = j) \\ 0, & \text{else} \end{cases} \quad (6.4)$$

In order to ensure that similar samples can have similar labels and the predicted labels on the labeled samples are consistent with the actual labels, the loss function F is defined as below:

$$\begin{aligned}
E(F) &= \frac{1}{2}\left(\sum_{i=1}^{m}\sum_{j=1}^{m}(W)_{ij}\left\|\frac{1}{\sqrt{d_i}}F_i - \frac{1}{\sqrt{d_j}}F_j\right\|^2\right) + \mu\sum_{i=1}^{l}\|F_i - Y_i\|^2 \\
&= \frac{1}{2}\sum_{i=1}^{m}\sum_{j=1}^{m}(W)_{ij}\left(\frac{1}{d_i}F_i^2 + \frac{1}{d_j}F_j^2 - 2\frac{1}{\sqrt{d_i d_j}}F_i F_j\right) + \mu\sum_{i=1}^{l}\|F_i - Y_i\|^2 \\
&= \frac{1}{2}\sum_{i=1}^{m}F_i^2 + \frac{1}{2}\sum_{j=1}^{m}F_j^2 - \sum_{i=1}^{m}SF_i^2 + \mu\sum_{i=1}^{l}\|F_i - Y_i\|^2 \quad (6.5) \\
&= \sum_{i=1}^{m}F_i^2 - \sum_{i=1}^{m}SF_i^2 + \mu\sum_{i=1}^{l}\|F_i - Y_i\|^2 \\
&= F^T(I - S)F + \mu\sum_{i=1}^{l}\|F_i - Y_i\|^2
\end{aligned}$$

where $S = D^{-\frac{1}{2}}WD^{-\frac{1}{2}}$ is the label propagation matrix, $D = diag(d_1, d_2, \ldots d_m)$ is a diagonal matrix composed of d_i, $d_i = \sum_{j=1}^{l+u}(W)_{ij}$ is the sum of the elements in the i-th row of matrix W, and μ is the regularization parameter. The first term ensures that the labels of similar samples are similar, and the second term makes the predicted results on the labeled samples are consistent with the real results.

Let

$$\frac{\partial E(F)}{\partial F} = \left[(I-S)+(I-S)^T\right]F + 2\mu(F-Y) = 0 \tag{6.6}$$

Then

$$F = (1-\alpha)(I-\alpha S)^{-1} Y \tag{6.7}$$

where $\alpha = \dfrac{1}{1+\mu}$. By substituting Y, the label prediction matrix F of unlabeled samples can be obtained accordingly.

6.2.2 Improved Random Forest Classification Algorithm

The conventional random forest algorithm is a supervised learning algorithm, but most of the samples obtained in actual engineering applications are unlabeled. Therefore, how to include a large number of unlabeled samples in the random forest learning is the main aim of this chapter. To solve this problem, an improved random forest algorithm is proposed and the implementation process of the algorithm is as follows:

Step 1: Randomly extract $n(n<l)$ samples from the data set D_u to form the data set D_u^1 and use the graph-based semi-supervised learning method described in the previous section to perform label prediction on this part of the data.

Step 2: Perform feature extraction and feature subset selection on the data set $D = \{D_l \cup D_u^1\}$ and select the time-domain features as the splitting attributes in the growth process of the decision tree considering the characteristics of the rotating machinery. Conduct k random samplings from the feature set T and extract m features each time to combine into a feature subset.

Step 3: For the data set $D = \{D_l \cup D_u^1\}$, perform k random sampling with replacement according to the bootstrap resampling method to generate k bootstrap sub-sample sets.

Step 4: For each bootstrap sub-sample set, build and train the decision trees accordingly.

Step 5: Use k bootstrap sub-sample sets to grow k decision trees and output the classification results of each decision tree for each data in the data set D_u^1.

Step 6: Compare the classification result of step 5 with the label prediction result of step 1, select the data with the same label and add it into the original labeled sample set D_l to form a new labeled sample set D_l'.

Step 7: For the new labeled data set D_l', follow steps 3 to 5 to grow k decision trees. Use the constructed k decision trees as the base classifier group of the random forest, and divide the features into different fault categories. Then the classification results of the k decision trees are voted to obtain the classification results of the random forest and realize the fault diagnosis.

6.3 EXPERIMENTAL VERIFICATION

To verify the prediction accuracy of the improved random forest algorithm, the data of normal gear and three types of faulty gears (worn, root crack, and tooth chipped) are collected during the experiments. In the wind turbine drivetrain diagnostic simulator, select six working conditions with rotation frequencies of 6 Hz and 10 Hz and load voltages of 8 V, 5 V, and 3 V. The data of the gearbox are collected from the vibration sensor and the torque sensor. Each type of data contains a small number of labeled samples and a large number of unlabeled samples.

The implementation of fault diagnosis of rotating machinery based on the improved random forest algorithm involves the selection of two parameters, which are the number of optimal decision trees $ntree$ and the optimal feature subset m. The value of $ntree$ is the number of bootstrap resampling in the experiment, which is determined according to the out-of-bag error during the experiment. For the new labeled samples D_l', the temporal feature extraction is performed according to the feature extraction method introduced in Section 6.2.2. The data after the feature extraction and the corresponding category labels $Y = \{y_1, y_2, y_3, y_4\}$ are used as the inputs of the random forest. As the number of decision trees increases, the error rate of the model is also changing and the overall trend of the error rate is decreasing. When the prediction accuracy of the model approaches the minimum and remains basically constant, the $ntree$ value at this time is selected as the number of optimal decision trees in the random forest model. In this experiment, the number of optimal decision trees $ntree$ is 480. The fault diagnosis results based on the improved random forest model using the updated labeled samples D_l' are shown in Table 6.1.

From the results given in Table 6.1 it is clear that the improved random forest algorithm has a relatively high classification accuracy for each type of fault. It can be seen from the confusion matrix in Table 6.1 that the algorithm has the best classification effect on y4, and the overall fault diagnosis accuracy can reach 90.2–1%.

TABLE 6.1
Confusion Matrix of Improved Random Forest Model

Actual	Predicted				Error rate
	y1	y2	y3	y4	
y1	109	1	10	0	0.092
y2	2	106	12	0	0.117
y3	6	16	98	0	0.183
y4	0	0	0	120	0

BIBLIOGRAPHY

Breiman, L. 1996. Bagging predictors. *Machine Learning*, 24, 123–140.
Breiman, L. 2001. Random forests. *Machine Learning*, 45, 5–32.
Chen, X.-W. & Liu, M. 2005. Prediction of protein–protein interactions using random decision forest framework. *Bioinformatics*, 21, 4394–4400.
Ding, Y., Yan, S., Zhang, Y., Dai, W. & Dong, L. 2016. Predicting the attributes of social network users using a graph-based machine learning method. *Computer Communications*, 73, 3–11.
Ho, T. K. 1998. The random subspace method for constructing decision forests. *IEEE Transactions on Pattern Analysis and Machine Intelligence*, 20, 832–844.
Huang, G., Song, S., Gupta, J. N. & Wu, C. 2014. Semi-supervised and unsupervised extreme learning machines. *IEEE Transactions on Cybernetics*, 44, 2405–2417.
Kilinc, O. & Uysal, I. 2018. GAR: An efficient and scalable graph-based activity regularization for semi-supervised learning. *Neurocomputing*, 296, 46–54.
Konerman, M. A., Beste, L. A., Van, T., Liu, B., Zhang, X., Zhu, J., Saini, S. D., Su, G. L., Nallamothu, B. K. & Ioannou, G. N. 2019. Machine learning models to predict disease progression among veterans with hepatitis C virus. *PloS one*, 14, e0208141.
Lee, S. L. A., Kouzani, A. Z. & Hu, E. J. 2010. Random forest based lung nodule classification aided by clustering. *Computerized Medical Imaging and Graphics*, 34, 535–542.
Lei, Y., He, Z. & Zi, Y. 2008. Fault diagnosis based on novel hybrid intelligent model. *Chinese Journal of Mechanical Engineering*, 44, 112–117.
Li, C., Zhu, J. & Zhang, B. 2017. Max-margin deep generative models for (semi-) supervised learning. *IEEE Transactions on Pattern Analysis and Machine Intelligence*, 40, 2762–2775.
Li, Y., Fang, Y. & Fang, J. 2011. Predicting residue–residue contacts using random forest models. *Bioinformatics*, 27, 3379–3384.
Ließ, M., Glaser, B. & Huwe, B. 2012. Uncertainty in the spatial prediction of soil texture: Comparison of regression tree and Random Forest models. *Geoderma*, 170, 70–79.
Liu, T., Abd-Elrahman, A., Morton, J. & Wilhelm, V. L. 2018a. Comparing fully convolutional networks, random forest, support vector machine, and patch-based deep convolutional neural networks for object-based wetland mapping using images from small unmanned aircraft system. *GIScience & Remote Sensing*, 55, 243–264.
Liu, W., Wang, J. & Chang, S.-F. 2012. Robust and scalable graph-based semisupervised learning. *Proceedings of the IEEE*, 100, 2624–2638.
Liu, X., Pan, S., Hao, Z. & Lin, Z. 2014. Graph-based semi-supervised learning by mixed label propagation with a soft constraint. *Information Sciences*, 277, 327–337.
Liu, Y., Xu, Z. & Li, C. 2018b. Distributed online semi-supervised support vector machine. *Information Sciences*, 466, 236–257.
Noi, P. T. & Kappas, M. 2018. Comparison of random forest, k-nearest neighbor, and support vector machine classifiers for land cover classification using Sentinel-2 imagery. *Sensors*, 18, 18.
Parkhurst, D. F., Brenner, K. P., Dufour, A. P. & Wymer, L. J. 2005. Indicator bacteria at five swimming beaches—analysis using random forests. *Water Research*, 39, 1354–1360.
Smith, A., Sterba-Boatwright, B. & Mott, J. 2010. Novel application of a statistical technique, Random Forests, in a bacterial source tracking study. *Water Research*, 44, 4067–4076.
Strobl, C., Malley, J. & Tutz, G. 2009. An introduction to recursive partitioning: Rationale, application, and characteristics of classification and regression trees, bagging, and random forests. *Psychological Methods*, 14, 323–348.
Waljee, A. K., Liu, B., Sauder, K., Zhu, J., Govani, S. M., Stidham, R. W. & Higgins, P. D. 2018. Predicting corticosteroid-free biologic remission with vedolizumab in Crohn's disease. *Inflammatory Bowel Diseases*, 24, 1185–1192.

Waljee, A. K., Sauder, K., Patel, A., Segar, S., Liu, B., Zhang, Y., Zhu, J., Stidham, R. W., Balis, U. & Higgins, P. D. 2017. Machine learning algorithms for objective remission and clinical outcomes with thiopurines. *Journal of Crohn's and Colitis*, 11, 801–810.

Yi, Y., Qiao, S., Zhou, W., Zheng, C., Liu, Q. & Wang, J. 2018. Adaptive multiple graph regularized semi-supervised extreme learning machine. *Soft Computing*, 22, 3545–3562.

Yu, C., Tian, C., Cheng, X., Qin, Y. & Shang, L. 2017. Multi-view collaborative semi-supervised classification algorithm based on diversity measurers of classifier with the combination of agreement and disagreement label rules. *Transactions of the Institute of Measurement and Control*, 39, 625–634.

Zhao, Y., Ball, R., Mosesian, J., De Palma, J.-F. & Lehman, B. 2014. Graph-based semi-supervised learning for fault detection and classification in solar photovoltaic arrays. *IEEE Transactions on Power Electronics*, 30, 2848–2858.

Zhu, X., Xiong, J. & Liang, Q. 2018. Fault diagnosis of rotation machinery based on support vector machine optimized by quantum genetic algorithm. *IEEE Access*, 6, 33583–33588.

7 Imbalanced Data Fault Diagnosis Based on Hybrid Feature Dimensionality Reduction and Varied Density-Based Safe-Level Synthetic Minority Oversampling Technique

7.1 INTRODUCTION

The complexity of unbalanced data distributions is an important reason for the difficulty in unbalanced data classification. For high-dimensional unbalanced data, the data distribution of minority class samples is sparse compared with majority class samples, and the high-dimensional features also include irrelevant or redundant features. These factors make the classification of high-dimensional imbalanced data very challenging.

At present, for the problem of imbalanced data classification, most studies focus on solving the problems caused by the imbalance between classes, but ignore the impact of imbalance within the class on data classification. For a certain class of sample data, small disjunction is one possible root cause of imbalance within the class. These small disjunctions are sample subsets of the same type which are not adjacent to each other. They have different distribution densities and an unequal number of samples. For classification problems, these small disjunctions may cause the classifier to learn the classification rules of this class insufficiently. For some small disjunctions with a small number of samples, they may even be considered as noisy data by the classifier, which will eventually cause low classification accuracy.

Aiming at the problem of high-dimensional features in imbalanced data, this chapter starts from the feature level and proposes a hybrid feature dimensionality reduction algorithm that combines feature selection methods and feature transformation methods. First, the compensated distance evaluation technique (CDET) is used to perform sensitive feature selection for high-dimensional features, and then the kernel principal component analysis (KPCA) algorithm is used to perform feature transformation and dimensionality reduction on the data after feature selection.

Aiming at the problem of data imbalance within a class, how to effectively identify small disjunctions in imbalanced data is the key. The density-based spatial clustering of applications with noise (DBSCAN) algorithm can be used to identify the distribution of data through clustering. Since the clustering parameters such as the neighborhood distance threshold of sample *Eps* and the minimum number of samples in a cluster *MinPts* adopted by the DBSCAN algorithm are globally unified parameters, this algorithm is mainly suitable for clustering samples with uniform distribution density. When the DBSCAN algorithm clusters small disjunctions with different distribution densities in imbalanced data within a class, the clustering results are often not ideal as some small disjunct data may be treated as noise. Therefore, this chapter proposes the varied density-based safe-level synthetic minority oversampling technique (VDB-SLSMOTE) algorithm. First, by improving the DBSCAN algorithm, a varied density-based spatial clustering of applications with noise (VDBSCAN) algorithm is proposed which achieves variable density clustering of minority samples by selecting multiple values of *Eps*. In addition, the safety level of minority samples in the SLSMOTE algorithm is redefined according to the value of each cluster to ensure that the synthetic data tends to be in the area with high data distribution density.

Combining the above problems and their corresponding solutions, this chapter proposes an imbalanced data processing method that combines hybrid feature dimensionality reduction technique and VDB-SLSMOTE algorithm. This method deals with high-dimensional imbalanced data from the feature level and the data level respectively. Firstly, the hybrid feature dimensionality reduction technique is used to perform hybrid dimensionality reduction on high-dimensional imbalanced data. Secondly, the VDBSCAN algorithm is used to cluster the minority samples to identify the small disjunctions in the minority samples and eliminate the noise data. The SLSMOTE algorithm is used to generate new samples in the cluster to deal with the high-dimensional feature problem and the intra-class imbalance problem and obtain a relatively low-dimensional balanced data set. Finally, the algorithm proposed in this chapter is applied to the fault diagnosis of unbalanced rotating machine gear data, and the effectiveness of the proposed algorithm is verified through a comprehensive analysis of its classification results.

7.2 DESIGN OF HYBRID FEATURE DIMENSIONALITY REDUCTION ALGORITHM

In the high-dimensional imbalanced data, the data distribution is relatively scattered due to the limited number of samples in the minority class. When classifying the imbalanced data, the minority samples may be easily treated as noise data, which intensifies the imbalance property between different classes. In addition, some features in high-dimensional data are irrelevant and redundant, and the existence of these features increases the complexity of the algorithm and the difficulty of classifying imbalanced data. For this reason, this chapter proposes a hybrid feature dimensionality reduction technique. First, the CDET algorithm is used to perform feature

Imbalanced Data Fault Diagnosis

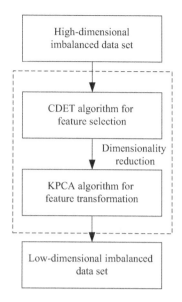

FIGURE 7.1 Flow chart of hybrid feature dimensionality reduction algorithm.

selection on the high-dimensional features in the imbalanced data set so that the features that are sensitive to classification are selected from these high-dimensional features. Then, the non-linear dimension reduction of data is conducted by the KPCA algorithm. Figure 7.1 shows the algorithm flow chart of hybrid feature dimensionality reduction.

7.2.1 Sensitive Feature Selection

In this chapter, the CDET algorithm is used to perform feature selection on high-dimensional imbalanced data. The CDET algorithm uses the inter-class distance and intra-class distance of the samples as the basic measurement criteria to define an evaluation factor that evaluates the sensitivity of each feature parameter. This evaluation factor can be selected to facilitate the classification of sensitive features. Suppose the original feature set is:

$$Q = \{q_{m,c,i}\} \tag{7.1}$$

where $m = 1, 2, \ldots, M_c$ represents the number of samples contained in each class, $c = 1, 2, \ldots, C$ represents the number of classes, $i = 1, 2, \ldots, M$ represents the number of features contained in each class, and $q_{m,c,i}$ represents the ith feature of the mth sample of the cth class. The specific steps of the CDET algorithm are shown in Table 7.1, where $\bar{\alpha}_i \in [0,1]$. By setting the value of $\bar{\alpha}_i$, the CDET algorithm will select features that are sensitive to the distance between classes and within classes according to this value. From Table 4.1, it can be found that the smaller the difference factor within a class, the more concentrated the distribution of samples of the same type is, which is suitable for classification; the greater the difference factor between classes,

TABLE 7.1
The Steps of the CDET Algorithm

Step	Operation	Equation
1	Calculate the average distance within the same class of samples	$d_{c,i} = \dfrac{\sum_{l,m=1}^{M_c} |q_{m,c,i} - q_{l,c,i}|}{M_c \times (M_c - 1)}, l \neq m$
2	Calculate the mean of the average distances within C' classes	$d_i^{(w)} = \dfrac{1}{C'} \sum_{c=1}^{C'} d_{c,i}$
3	Calculate the difference factor of the average distance within the class	$v_i^{(w)} = \dfrac{\max(d_{c,i})}{\min(d_{c,i})}$
4	Calculate the mean of each feature of each class	$u_{c,i} = \dfrac{\sum_{m=1}^{M_c} q_{m,c,i}}{M_c}$
5	Calculate the average distance between different classes	$d_i^{(b)} = \dfrac{\sum_{c,e=1}^{C'} |u_{e,i} - u_{c,i}|}{C' \times (C' - 1)}, c \neq e$
6	Calculate the difference factor of the distance between classes	$v_i^{(b)} = \dfrac{\max(|u_{e,i} - u_{c,i}|)}{\min(|u_{e,i} - u_{c,i}|)}, c \neq e$
7	Calculate the weighting factor	$\lambda_i = \dfrac{1}{\dfrac{v_i^{(w)}}{\max(v_i^{(w)})} + \dfrac{v_i^{(b)}}{\max(v_i^{(b)})}}$
8	Calculate distance evaluation factor	$\alpha_i = \lambda_i \dfrac{d_i^{(b)}}{d_i^{(w)}}$
9	Normalize distance evaluation factor	$\bar{\alpha}_i = \dfrac{\alpha_i}{\max(\alpha_i)}$

the more apparent the boundary between the samples of different classes is, which is also suitable for data classification.

7.2.2 Dimension Reduction of Features

The hybrid feature dimension reduction algorithm proposed in this chapter consists of feature selection and feature transformation. In the feature selection stage, the CDET algorithm is used for sensitive feature selection of high-dimensional features, and the KPCA algorithm is used in the feature transformation stage. The main idea of KPCA is to map the data set to the high-dimensional kernel space through the mapping function ϕ, and then use principal component analysis to reduce the data dimension in the high-dimensional kernel space. When performing feature dimension reduction on high-dimensional data using KPCA, it may not be feasible to compute the covariance matrix between features directly due to time complexity. Therefore, after feature selection on the original high-dimensional data using CDET, the processed data set is fed into the KPCA for computation. The implementation of the KPCA algorithm is as follows:

Imbalanced Data Fault Diagnosis

Assuming that the data set after CDET feature selection is $X = [x_1, x_2, \ldots, x_N]^T$, then the data set after the mapping function ϕ is $Z = [\phi(x_1), \phi(x_2), \ldots, \phi(x_N)]^T$, where N represents the size of the sample. The covariance matrix of Z can be obtained by the following equation:

$$\text{cov} = \frac{1}{N-1} \sum_{i=1}^{N} \phi(x_i) \phi(x_i)^T = \frac{1}{N-1} Z^T Z \tag{7.2}$$

Since the mapping function ϕ is unknown, the value of Z and its covariance cannot be directly obtained. Therefore, the Gaussian kernel function is introduced:

$$k(x_i, x_j) = \phi(x_i)^T \phi(x_j) = \left(-\frac{\|x_i - x_j\|}{2\sigma^2} \right) \tag{7.3}$$

The kernel matrix is:

$$K = \begin{bmatrix} k(x_1, x_1) & \cdots & k(x_1, x_N) \\ \vdots & \ddots & \vdots \\ k(x_N, x_1) & \cdots & k(x_N, x_N) \end{bmatrix} = ZZ^T \tag{7.4}$$

The eigenvalues λ and eigenvectors u of the kernel matrix K satisfy the following equation:

$$ZZ^T u = \lambda u \tag{7.5}$$

From Equations (7.2) and (7.4), it can be concluded that the matrices K and Cov have the same eigenvalues λ, and the eigenvector of Cov is $Z^T u$. By directly selecting the eigenvectors corresponding to the first k large eigenvalues to form the transformation matrix V, then the result after KPCA dimension reduction is:

$$D = V^T Z \tag{7.6}$$

7.3 DESIGN OF VARIED DENSITY-BASED SAFE-LEVEL SYNTHETIC MINORITY OVERSAMPLING TECHNIQUE

The main idea of the VDB-SLSMOTE algorithm is to implement variable density clustering of small disjunctions with uneven distribution density by using multiple *Eps* values through the VDBSCAN algorithm. Therefore, the small disjunctions can be effectively identified for the problem of intra-class imbalance. Then the SLSMOTE algorithm is improved based on the *Eps* values in the clusters to ensure the effectiveness of the data synthesized by the algorithm in the corresponding clusters to deal with the imbalance problems between classes and within classes.

The DBSCAN is a density-based clustering algorithm, which implements the clustering of data through parameters of *Eps* and *MinPts*. The *Eps* and *MinPts* in the

DBSCAN algorithm are global parameters; therefore, the algorithm can generate better clustering results when clustering data with uniform data distribution. But the clustering effect is limited when dealing with data of uneven distribution, and some samples with sparse data distribution may even be recognized as noise during clustering. Therefore, an improved algorithm based on the DBSCAN algorithm, VDBSCAN algorithm, is proposed in this chapter.

The VDBSCAN algorithm reflects the sparseness of the distribution density of the samples by calculating the distance between each sample and its nearest neighbor. The greater the distance between the sample and its nearest neighbor, the sparser the sample distribution. The sparse sample distribution means that the sample and its neighbors don't satisfy the requirement for forming a cluster. Therefore, a distance threshold is set to remove samples whose distance exceeds the threshold. In addition, the value of *DK* of the remaining *kdist* is computed, and then the value *kdist$_i$* of the sample corresponding to the point where the normalized *DK* value is greater than one is the distance breakpoint of *k*-distance curve. These values are defined as preselected *Eps* values. As *MinPts* indicates the minimum number of samples required to form a cluster, the distance between two adjacent preselected *Eps* values in the *k*-distance curve should be greater than *MinPts*. After removing the preselected *Eps* values that do not meet this condition, the remaining value is the final *Eps* value. Table 7.2 shows the specific steps of the VDB-SLSMOTE algorithm.

The VDB-SLSMOTE algorithm can help identify the small disjunctions in the imbalanced data within the class and introduce reasonable parameters when synthesizing the data. By avoiding the generation of noise data, this method can improve the effectiveness of the synthesis of data.

7.4 EXPERIMENT AND RESULTS

7.4.1 Data Classification Method

An imbalanced data classification method based on hybrid feature dimension reduction and VDB-SLSMOTE is proposed in this chapter. The overall algorithm flow chart of this chapter is shown in Figure 7.2, and the specific implementation steps are as follows:

Step 1: Use the CDET algorithm to select sensitive features that are useful for classification from high-dimensional features.

Step 2: Use the KPCA algorithm to perform feature transformation and dimension reduction on the data set after feature selection.

Step 3: Obtain the sorted *k*-distance curve and *DK* curve through computation and select multiple clustering parameters *Eps* according to the distance threshold.

Step 4: Sort the *Eps* values in ascending order and use the parameters *Eps$_i$* in turn to cluster the minority samples to obtain the clusters $C = \{C_1, C_2, ..., C_m\}$ and allocate the number of samples to be synthesized in each cluster.

Step 5: Use the improved SLSMOTE algorithm to generate new samples in each cluster.

TABLE 7.2
The Steps of the VDB-SLSMOTE Algorithm

Step	Operations
1	VDBSCAN clustering:

(1) Set values of k and *MinPts*.
(2) Calculate the Euclidean distance between all minority samples and their k nearest neighbors to obtain *kdist* and distance threshold *Threshold* and remove noise data according to *Threshold*.
(3) Remove the *kdist* value corresponding to the noise data and calculate the remaining first-order forward difference value *DK* of *kdist*.
(4) Standardize *DK* and use the $kdist_i$ value corresponding to the point whose *DK* value is greater than one after the standardization as the preselected *Eps* value.
(5) Choose the final $Eps = \{eps_1, eps_2, ..., eps_i\}$ according to the setting of *MinPts*, and the selection rule should satisfy: the number of samples between two adjacent preselected *Eps* values is greater than or equal to *MinPts*.
(6) Sort the selected final *Eps* values in ascending order and call the parameter values among them to cluster in order to obtain clusters $C = \{C_1, C_2, ..., C_m\}$.

2	Assign the number of samples to be synthesized for each cluster:

(1) Calculate the total number of samples needed to synthesize the minority class as $N_{maj} - N_{min}$.
(2) Calculate the number of minority samples in each cluster as NC_i, then the number of minority samples to be synthesized in each cluster is $Num_i = NC_i/(N_{maj} - N_{min})$.

3	Clustering with improved SLSMOTE algorithm:

(1) Replace the K value of the K nearest neighbor in the SLSMOTE algorithm with the eps_i value of the cluster when performing data synthesis in each cluster.
(1) Use the improved SLSMOTE algorithm to synthesize data in each cluster to obtain a relatively balanced data set.

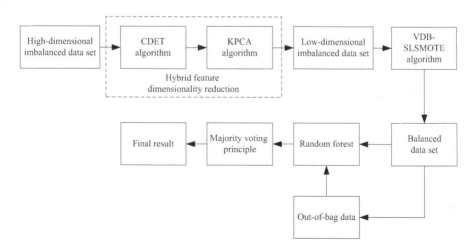

FIGURE 7.2 Overall algorithm flow chart.

Step 6: Use the Bagging algorithm to select multiple training subsets, and use the out-of-bag data as the test data.
Step 7: Train multiple decision trees according to the Gini index and combine them to form a random forest classifier.
Step 8: Put the test data samples into the random forest to obtain the final classification result.

7.4.2 Experiment Platform

To verify the effectiveness of the above-mentioned methods, the fault diagnosis experiments are conducted on the wind turbine drivetrain diagnosis simulator. The main components of the experiment platform are the following: four sensors to collect data in the experiment (two vibration sensors, one torque sensor, and one pressure sensor), drive motor, parallel shaft gearbox, planetary gearbox, and load brake. The multiple working conditions of the wind turbine system are achieved by setting different motor speeds and load voltages through the controller to simulate the actual working environment. In this experiment, nine different working conditions are formed by the combination of three motor frequencies and three load voltages. The specific settings of the working conditions are shown in Table 7.3.

7.4.3 Feature Extraction

Feature extraction can effectively extract important information in the original data set. By computing the features of the original data in the time domain, frequency domain, or time-frequency domain and constructing the feature vector to replace the original data, the characteristics of the data information can be reflected concisely and effectively. Table 7.4 lists the characteristic parameters in the time domain and frequency domain in the gearbox fault diagnosis on the experiment platform. In Table 7.4, $x(n)$ represents a certain segment of time-domain signal, $n = 1, 2, ..., N$ represents the number of samples, $s(k)$ represents the signal spectrum, $k = 1, 2, ..., K$ represents the number of spectral lines, f_k is the frequency value of the spectral line

TABLE 7.3
Setting of Working Conditions

Working Condition	Load Voltage	Rotation Frequency (Hz)
1		6
2	3 V	10
3		14
4		6
5	5 V	10
6		14
7		6
8	8 V	10
9		14

TABLE 7.4
Characteristic Parameters

Parameter	Mathematical Expression	Parameter	Mathematical Expression
F_1	$\dfrac{\sum_{n=1}^{N} x(n)}{N'}$	F_{10}	$\dfrac{F_4}{\dfrac{1}{N'}\sum_{n=1}^{N'} |x(n)|}$
F_2	$\sqrt{\dfrac{\sum_{n=1}^{N'}(x(n)-F_1)^2}{N'-1}}$	F_{11}	$\dfrac{F_5}{\dfrac{1}{N'}\sum_{n=1}^{N'} |x(n)|}$
F_3	$\left(\dfrac{\sum_{n=1}^{N'} |x(n)|}{N}\right)^2$	F_{12}	$\dfrac{\sum_{k=1}^{K} s(k)}{K}$
F_4	$\sqrt{\dfrac{\sum_{n=1}^{N'}(x(n))^2}{N'}}$	F_{13}	$\dfrac{\sum_{k=1}^{K} f_k s(k)}{\sum_{k=1}^{K} s(k)}$
F_5	$\max|x(n)|$	F_{14}	$\dfrac{\sum_{k=1}^{K} f_k^2 s(k)}{\sum_{k=1}^{K} s(k)}$
F_6	$\dfrac{\sum_{n=1}^{N'}(x(n)-F_1)^3}{(N'-1)(F_2)^3}$	F_{15}	$\sqrt{\dfrac{\sum_{k=1}^{K} f_k^2 s(k)}{\sum_{k=1}^{K} s(k)}}$
F_7	$\dfrac{\sum_{n=1}^{N'}(x(n)-F_1)^4}{(N'-1)(F_2)^4}$	F_{16}	$\dfrac{\sum_{k=1}^{K}(f_k-F_{13})^2 s(k)}{K}$
F_8	$\dfrac{F_5}{F_4}$	F_{17}	$\sqrt{\dfrac{\sum_{k=1}^{K}(f_k-F_{13})^2 s(k)}{K}}$
F_9	$\dfrac{F_5}{F_3}$		

k, the parameters $F_1 \sim F_{11}$ are time domain features, and the parameters $F_{12} \sim F_{17}$ are frequency domain features.

7.4.4 Data Acquisition

Five different planetary gearbox faults are selected in the experiment, and the sampling frequency in the experiment is set to 5120 Hz with a sampling time of 6.4 s. In this chapter, a sliding time window is used for feature extraction, where the length of the time window is 1024 data points and the step size is 256 data points. So there are 1125 samples obtained for each type of gear. There are a total of four sensors in the experiment platform and 17 features are extracted from each sensor; therefore, there are a total of 68 features in each sample. To construct an imbalanced data set, the random under-sampling is conducted in the five types of gear samples. Table 7.5 shows the types of gears and the collected number of samples.

TABLE 7.5
Gear Types and Number of Samples

Serial Number	Gear Type	Number of Samples
1	Normal	900
2	Missing tooth	630
3	Root crack	450
4	Chipped tooth	270
5	Surface worn	180

7.4.5 Results Analysis

In order to verify the effectiveness of the proposed algorithm proposed in this chapter, the proposed algorithm is applied to the fault diagnosis of the planetary gearbox in the experiment. The parameter selection of the Gaussian kernel function in the KPCA algorithm is $\sigma = 8$ and $k = MinPts = 9$ in the VDBSCAN algorithm. Table 7.6 shows the multi-category evaluation index of imbalanced data, where $macro\_P$ reflects the average precision, $macro\_R$ reflects the average recall rate, $macro\_F1$ reflects average $F1$ metric, $Kappa$ reflects the overall classification effect of the random forest on unbalanced gear fault data. From these index values, it can be seen that after the processing of the algorithm proposed in this chapter, the classification accuracies using random forest are high in all classes.

Table 7.7 shows the confusion matrix for multiple classifications. It can be seen that the classifier has high classification accuracy for each class in the gearbox fault

TABLE 7.6
Multi-Class Evaluation Index

Data set	macro_P	macro_R	macro_F1	Kappa
Gearbox fault data	0.9991	0.9991	0.9991	0.9989

TABLE 7.7
Confusion Matrix of Improved Random Forest Model

Actual	Predicted					Error rate
	1	2	3	4	5	
1	697	0	0	0	0	0.000
2	1	697	1	0	0	0.003
3	0	1	702	0	0	0.001
4	0	0	0	695	0	0.000
5	0	0	0	0	697	0.000

data. Based on the above experimental results, it can be concluded that the hybrid feature extraction and VDB-SLSMOTE algorithm proposed in this chapter can effectively solve the high-dimensional feature and intra-class imbalance problems in the imbalanced multi-class data classification.

BIBLIOGRAPHY

Breiman, L. 2001. Random forests. *Machine Learning*, 45, 5–32.
Buda, M., Maki, A. & Mazurowski, M. A. 2018. A systematic study of the class imbalance problem in convolutional neural networks. *Neural Networks*, 106, 249–259.
Bunkhumpornpat, C., Sinapiromsaran, K. & Lursinsap, C. Safe-level-smote: Safe-level-synthetic minority over-sampling technique for handling the class imbalanced problem. *Pacific-Asia Conference on Knowledge Discovery and Data Mining*, 2009. 475–482.
Bunkhumpornpat, C., Sinapiromsaran, K. & Lursinsap, C. 2012. DBSMOTE: density-based synthetic minority over-sampling technique. *Applied Intelligence*, 36, 664–684.
Chawla, N. V., Bowyer, K. W., Hall, L. O. & Kegelmeyer, W. P. 2002. SMOTE: synthetic minority over-sampling technique. *Journal of Artificial Intelligence Research*, 16, 321–357.
Ester, M., Kriegel, H.-P., Sander, J. & Xu, X. A density-based algorithm for discovering clusters in large spatial databases with noise. *Second International Conference on Knowledge Discovery and Data Mining*, 1996. 226–231.
Gan, D., Shen, J., An, B., Xu, M. & Liu, N. 2020. Integrating TANBN with cost sensitive classification algorithm for imbalanced data in medical diagnosis. *Computers & Industrial Engineering*, 140, 106266.
Guo, H., Li, Y., Shang, J., Gu, M., Huang, Y. & Gong, B. 2017. Learning from class-imbalanced data: Review of methods and applications. *Expert Systems with Applications*, 73, 220–239.
He, H. & Garcia, E. A. 2009. Learning from imbalanced data. *IEEE Transactions on Knowledge and Data Engineering*, 21, 1263–1284.
Imam, T., Ting, K. M. & Kamruzzaman, J. z-SVM: An SVM for improved classification of imbalanced data. *Australasian Joint Conference on Artificial Intelligence*, 2006. Springer, 264–273.
Krawczyk, B. 2016. Learning from imbalanced data: open challenges and future directions. *Progress in Artificial Intelligence*, 5, 221–232.
Lei, Y., He, Z. & Zi, Y. 2008. Fault diagnosis based on novel hybrid intelligent model. *Chinese Journal of Mechanical Engineering*, 44, 112–117.
Li, K., Xie, P., Zhai, J. & Liu, W. An improved adaboost algorithm for imbalanced data based on weighted knn. *2017 IEEE 2nd International Conference on Big Data Analysis (ICBDA)*, 2017. IEEE, 30–34.
Li, X., Chen, Z. & Yang, F. Exploring of clustering algorithm on class-imbalanced data. *2013 8th International Conference on Computer Science & Education*, 2013. 89–93.
Li, Y., Guo, H., Liu, X., Li, Y. & Li, J. 2016. Adapted ensemble classification algorithm based on multiple classifier system and feature selection for classifying multi-class imbalanced data. *Knowledge-Based Systems*, 94, 88–104.
Moayedikia, A., Ong, K.-L., Boo, Y. L., Yeoh, W. G. & Jensen, R. 2017. Feature selection for high dimensional imbalanced class data using harmony search. *Engineering Applications of Artificial Intelligence*, 57, 38–49.
Shahee, S. A. & Ananthakumar, U. 2020. An effective distance based feature selection approach for imbalanced data. *Applied Intelligence*, 50, 717–745.
Sun, Z., Song, Q., Zhu, X., Sun, H., Xu, B. & Zhou, Y. 2015. A novel ensemble method for classifying imbalanced data. *Pattern Recognition*, 48, 1623–1637.

Wang, Y.-R., Sun, G.-D. & Jin, Q. 2020. Imbalanced sample fault diagnosis of rotating machinery using conditional variational auto-encoder generative adversarial network. *Applied Soft Computing*, 92, 106333.

Yin, S., Zhu, X. & Jing, C. 2014. Fault detection based on a robust one class support vector machine. *Neurocomputing*, 145, 263–268.

Index

A

acceleration, 55
accelerometer, 21, 44
adaboost, 77
adaptation, 39, 40, 46
adversarial, 5, 8, 9, 39–46, 78
auto-encoder, 46, 78
Automation, 25

B

backpropagation, 51
backward, 52
bagging, 5, 7, 9, 37, 50, 64, 74
bearing, 1, 9, 11, 12, 20–22, 24, 25, 31, 39, 40, 42, 44–46, 57
bootstrap, 5, 7, 29, 30, 33, 36, 51, 62, 63
Butterworth-Heinemann, 24

C

classification, 3, 5–9, 11, 15, 20, 22, 24, 27, 29–31, 34–37, 40, 42, 43, 45–52, 55–57, 59, 60, 62–65, 67–70, 72, 74, 76, 77
class-imbalanced, 77
clustering, 7–9, 37, 46, 64, 68, 71–73, 77
CNN, 12, 14, 15, 24, 57
co-adaptation, 18
convergence, 28
convolution, 3, 4, 9, 11–19, 21–24, 42–44, 46–49, 51–53, 55, 57, 64, 77
cross-entropy, 50

D

data-driven, 19
DBSCAN, 9, 68, 71, 72
DBSMOTE, 77
density-based, 7–9, 67, 68, 71, 77
density-reachable, 8
detection, 1, 9, 19, 23–25, 28, 46, 65, 78
diagnosis, 1–3, 10–13, 19–25, 27–37, 39, 40, 42–48, 51, 54–57, 59, 60, 63–65, 67, 68, 74, 76–78
dimensionality, 7, 15, 32, 57, 61, 67–73, 77
discrepancy, 39–45
discriminator, 5, 40, 41, 44
disjunction, 67, 68, 71, 72
distance, 7, 40, 41, 44, 60, 67–70, 72, 73, 77
distribution, 5, 32, 39–44, 49, 50, 64, 67–69, 71, 72
disturbances, 11
divergence, 40
domain, 5, 12, 27, 29, 32, 33, 39, 40, 42–46, 74, 75
down-sampling, 3, 51
dropout, 14, 18, 19, 23

E

eigenvalues, 71
eigenvectors, 4, 30, 71
embedding, 22, 42
entropy, 43, 46
Euclidean, 7, 60, 73

F

fault, 1–3, 9–12, 19–25, 27–37, 39, 40, 42–48, 51, 55–57, 59–61, 63–65, 67, 68, 74–78
feature, 1–7, 9, 11–13, 22–24, 27–30, 32, 33, 35, 36, 39, 40, 42–51, 55, 59, 62, 63, 67–75, 77
forest, 6, 7, 9, 10, 27–30, 33–37, 47, 48, 50, 51, 55–57, 59, 60, 62–64, 73, 74, 76, 77
fusion, 11

G

Gaussian, 60, 61, 71, 76
gearbox, 1, 23–25, 27–33, 37, 47, 48, 54–57, 59, 63, 68, 74–76
generalization, 5, 9
generative, 5, 8, 9, 39–46, 59, 64, 77, 78
gradient, 2, 3, 5, 9, 16–18, 20, 40, 41, 54
graph-based, 59–62, 64, 65

H

high-dimensional, 4, 22, 35, 67–70, 72, 77
high-efficiency, 48
hyperbolic, 3, 16
hyperplane, 35

I

imbalanced, 8, 9, 46, 67–69, 71–73, 75–78
inter-class, 69
interpolation, 8
intra-class, 68, 69, 71, 77
iteration, 5, 20, 22, 23, 51, 54

79

J

Jensen-Shannon, 40

K

Kantorovich–Rubinstein, 41
kdist, 7, 72, 73
k-distance, 7, 72
kernel, 3, 4, 13, 14, 18, 22, 41–43, 49, 52, 53, 68, 70, 71, 76
k-nearest, 8, 64, 77
Knowledge-Based, 46, 77

L

learning-based, 2, 42
loss, 2, 15, 16, 18, 22, 27, 43, 44, 50, 52–54, 61
low-dimensional, 22, 68, 69, 73
LSTM, 25

M

model-based, 23, 25, 57, 59
multi-category, 76
multi-class, 76, 77
multi-classification, 15, 35, 36, 49
multi-condition, 12, 20
multi-dimensional, 32
multi-domain, 39
multi-layer, 3–5, 9, 42
multi-level, 3
multi-type, 47

N

neighbor, 7, 8, 22, 64, 68, 72, 73
network, 1–5, 8–25, 37, 39–52, 54, 55, 57, 64, 77, 78
network-based, 11, 25, 28
neural, 1–5, 9–19, 21–25, 28, 37, 42, 43, 46–49, 55, 57, 64, 77
neural-network-based, 24
neuron, 2–4, 15, 17–19, 22, 49, 50, 53
non-differentiable, 40
non-leaf, 6
nonlinear, 2, 4, 13, 25, 35, 69
non-parametric, 41

O

one-dimensional, 12, 13, 46–49, 55, 57
optimal, 5, 6, 23, 29, 30, 33, 35, 40–42, 51, 54, 60, 63, 65
out-of-bag, 6, 30, 33, 63, 73, 74
over-complete, 45
overfitting, 3, 18, 19, 27, 28, 48–50
over-sampling, 8, 9, 67, 68, 71, 77

P

pooling, 3, 12, 13, 15, 18, 22, 42, 49, 51–53
precision, 76
probability, 6, 15, 18, 19, 22, 41, 43, 49, 50, 55
propagation, 2, 16, 46, 51–53, 60, 61, 64

R

real-time, 1, 24
recurrent, 4, 5, 11–17, 19, 21–24
regression, 6, 7, 9, 10, 29, 37, 50, 51, 64
regularization, 62, 64
resampling, 5, 7, 29, 30, 33, 36, 62, 63
RNN, 4
rolling, 20, 24, 42, 45, 46
rotating, 11, 19, 23–25, 27–33, 36, 39, 47, 48, 55–57, 59, 60, 62, 63, 65, 68, 74, 78

S

safe-level, 8, 9, 67, 68, 71, 77
semi-supervised, 59, 60, 62, 64, 65
SLSMOTE, 68, 71–73
softmax, 14–16, 42, 43, 49, 50, 56
sub-networks, 18
sub-sample, 5–7, 62

T

t-distributed, 22
time-direction, 13
time-domain, 27, 28, 32, 48, 62, 74
time-frequency, 74
transfer, 5, 39, 40, 42, 45, 46
turbine, 24, 27, 31, 37, 54, 55, 63, 74

U

unbalanced, 67, 68, 76
under-sampling, 75
unlabeled, 39, 40, 46, 59, 60, 62, 63
unsupervised, 24, 40, 46, 64

V

vanishing, 40, 41
VDB-High-dimensional, 73
VDBSCAN, 68, 71–73, 76
VDB-SLSMOTE, 68, 71–73, 77
vibration, 11, 12, 25, 28, 30–32, 44, 47, 56, 57, 63, 74
visualization, 22, 46

W

Wasserstein, 8, 9, 39–42, 44–46